D1364517

# Contemporary Biographies in Physics

# Contemporary Biographies in Physics

SALEM PRESS

A Division of EBSCO Publishing

Ipswich, Massachusetts

**GREY HOUSE PUBLISHING**

Copyright © 2013, by Salem Press, A Division of EBSCO Publishing, Inc. All rights reserved. No part of this work may be used or reproduced in any manner whatsoever or transmitted in any form or by any means, electronic or mechanical, including photocopy, recording, or any information storage and retrieval system, without written permission from the copyright owner. For permissions requests, contact proprietarypublishing@ebscohost.com.

*Contemporary Biographies in Physics,* 2013, published by Grey House Publishing, Inc., Amenia, NY, under exclusive license from EBSCO Publishing, Inc.

∞ The paper used in these volumes conforms to the American National Standard for Permanence of Paper for Printed Library Materials, Z39.48-1992 (R1997).

**Library of Congress Cataloging-in-Publication Data**

Contemporary biographies in physics.
    pages cm
  Includes bibliographical references and indexes.
  ISBN 978-1-58765-996-6 (hardcover)
  1. Physicists--Biography. 2. Women physicists--Biography.
  QC15.C66 2013
  530.092'2--dc23

                                                            2012044527


ebook ISBN: 978-1-58765-998-0

PRINTED IN THE UNITED STATES OF AMERICA

# Contents

# Publisher's Note

*Contemporary Biographies in Physics* is a collection of thirty-one biographical sketches of "living leaders" in the fields of physics. All of these articles come from the pages of *Current Biography*, the monthly magazine renowned for its unfailing accuracy, insightful selection, and the wide scope of influence of its subjects. These up-to-date profiles draw from a variety of sources and are an invaluable resource for researchers, teachers, students, and librarians. Students will gain a better understanding of the educational development and career pathways—the rigors and rewards of a life in science—of the contemporary scientist to better prepare themselves for a scientific career.

The geographical scope of *Contemporary Biographies in Physics* is broad; selections span the Eastern and Western Hemispheres, covering numerous major geographical and cultural regions. While most of the figures profiled are practicing scientists in their respective fields, the selection also includes scientifically trained government officials, institutional directors, and other policy leaders who are helping to shape the future of science by setting agendas and advancing research. Scientific fields covered range from biophysics and astrophysics, to electronics and mechanical engineering, to climatology, meteorology, and oceanography.

Articles in *Contemporary Biographies in Physics* range in length from roughly 1,000 to 4,000 words and follow a standard format. All articles begin with ready-reference listings that include birth details and concise identifications. The articles then generally divide into several parts, including the Early Life and Education section, which provides facts about the scientists' early lives and the environments in which they were reared, as well as their educational background; and Life's Work, a core section that provides straightforward accounts of the periods in which the profiled subjects made their most significant contributions to science. Often, a final section, Significance, provides an overview of the scientists' places in history and their contemporary importance. Essays are supplemented by bibliographies, which provide starting points for further research.

As with other Salem Press biographical reference works, these articles combine breadth of coverage with a format that offers users quick access to the particular information needed. For convenience of reference, articles are arranged alphabetically by scientists' names, and an appendix lists scientists' names by their country of origin. In addition, a general bibliography offers a comprehensive list of works for students seeking out more information on a particular scientist or subject, while a separate bibliography of selected works highlights the significant published works of the scientists profiled. An appendix consisting of ten historical biographies of "Great Physicists," culled from the Salem Press *Great Lives* series, introduces readers to scientists of historical significance integral to the genesis of physics and whose work and research revolutionized science.

The editors of Salem Press wish to extend their appreciation to all those involved in the development and production of this work; without their expert contribution, projects of this nature would not be possible. A full list of contributors appears at the beginning of this volume.

# Contributors List

Mike Batistick
Matt Broadus
Damitri Cavalli
Andrew I. Cavin
In-Young Chang
Forrest Cole
Matthew Creamer
Christopher Cullen
Jennifer Curry
Kathleen A. D'Angelo
Sara J. Donnelly
Karen E. Duda
William Dvorak
Ronald Eniclerico
Karen E. Euda
Kaitlen J. Exum
Dan Firrincili
Terence J. Fitzgerald
Willie Gin
Peter G. Herman
Josha Hill
Martha A. Hostetter
Louisa Jennings
Virginia Kay
Patrick Kelly
David J. Kim
Dmitry Kiper
Christopher Luna
Nicholas W. Malinowski
Christopher Mari
Paul B. McCaffrey

Margaret R. Mead
David Moldawer
Majid Mozaffari
Bertha Muteba
Tracy O'Neill
Geoff Orens
Joanna Padovano
Yongsoo Park
Kenneth J. Partridge
Jamie E. Peck
Constantina Petropoulos
David Ramm
Mari Rich
Josh Robertson
Gregory K. Robinson
Albert Rolls
Margaret E. Roush
Liliana Segura
Olivia Jane Smith
Brian Solomon
Claire Stanford
Luke A. Stanton
Kate Stern
Maria A. Suarez
Hope Tarullo
Aaron Tassano
Cullen F. Thomas
Hallie Rose Waxman
Lara Weibgen
Selma Yampolsky

# Contemporary Biographies in Physics

# Abrikosov, Alexei

Russian physicist

---

**Born**: June 25, 1928; Moscow, Russia

---

On October 8, 2003, Alexei A. Abrikosov, along with Vitaly L. Ginzburg and Anthony J. Leggett, was awarded the Nobel Prize in Physics for his pioneering contributions to the theory of superconductivity. He earned the award because he had explained theoretically how "Type-II superconductors allow superconductivity and magnetism to exist at the same time and remain superconductive in high magnetic fields," as the press release from the Nobel committee read. In addition to the Nobel Prize, he has been honored with the Lenin Prize (1966), the Fritz London Award (1972), the USSR State Prize (1982), the USSR Academy of Sciences' Landau Award (1989), and the International John Bardeen Award (1991). Abrikosov has also received France's Honorary Citizenship of Saint Emilion. In 1975 he received an honorary doctorate from the University of Lausanne, in Switzerland. In 2001, he was elected to the Royal Society of London as a foreign member.

## Early Life and Education

Alexei Abrikosov was born on June 25, 1928, in Moscow, Russia. After graduating from Moscow State University in 1948, Abrikosov was admitted to the Kapitsa Institute for Physical Problems in Moscow. In 1951 he earned a doctorate in physics from the Institute on the basis of his dissertation, which explored the theory of thermal diffusion in plasma. Four years later, Abrikosov received his second doctorate in physics from the Institute for his dissertation on quantum electrodynamics at high energies.

## Life's Work

As a research associate with the Kapitsa Institute, Abrikosov began conducting research into the phenomenon of superconductivity. In

1911 the Dutch physicist Heike Kamerlingh Onnes (1853–1926) discovered superconductivity by cooling mercury to a few degrees above absolute zero and observing that electrical resistance in the metal disappeared. Kamerlingh Onnes's discovery earned him the Nobel Prize in Physics in 1913. In 1950 two Russian physicists, Vitaly Ginzburg and Lev Landau, published a scientific paper that offered a theory of how superconductivity worked. Ginzburg and Landau devised mathematical equations that explained why superconductivity and magnetism could coexist in some superconducting materials, but not others. (Landau won the Nobel Prize in Physics in 1962. Ginzburg shared the Nobel Prize in Physics in 2003 with Abrikosov and Anthony Leggett, a British physicist.) Abrikosov's colleague and roommate at the Institute, Nikolay Zavaritskii, began measuring the critical magnetic field of thin superconducting films to see if Ginzburg and Landau were correct in predicting their behavior. "[Ginzburg's and Landau's theory] and experiment fitted perfectly, including the change of the nature of the transition: first order at larger thickness and second order at smaller ones," Abrikosov recalled in his Nobel lecture, as posted on the Nobel e-Museum website. Zavaritskii's supervisor, Alexander Salnikov, was not satisfied with the results, however, because the young physicist had used films that were prepared at room temperature. "The atoms of the metal, evaporated on a glass substrate, could agglomerate, and there the film actually consisted of small droplets," Abrikosov recalled in his lecture. "In order to avoid that, Salnikov recommended to maintain the glass substrate at helium temperature during evaporation and until the measurements were finished. Then every metal atom hitting the surface would stick to its place, and the film would be homogeneous." When he tried the experiment again, following Salnikov's instructions, Zavaritskii found that the results did not confirm Ginzburg's and Landau's predictions. "Discussing these results with Zavaritskii, we couldn't believe that the theory was wrong: it was so beautiful, and fitted so well [with] the previous data," Abrikosov said in his lecture. "Therefore we tried to find some solution in the framework of the theory itself." Abrikosov found that the Ginzburg-Landau parameter, which formed the basis of the two physicists' equations, had small values because they were calculated from the surface energy between the normal and superconducting phases of the superconductor. When the value of the parameter was increased, the surface energy between the normal and superconducting phases became negative. Ginzburg and Landau kept the value of their parameter small because the existence of negative surface energy contra-

> **"With his theory of vortices, Abrikosov invented a whole new discipline."**

dicted the existence of the intermediate state in a superconductor. Abrikosov experimented with negative surface energy and discovered that the transition was of the second order for superconducting films of any thickness. He concluded that a special type of superconductors existed, which he and his colleagues called superconductors of the second group. These eventually became known as Type-II superconductors. Ginzburg and Landau had used Type-I superconductors, which expel magnetic fields, in their experiments. By contrast, Type-II superconductors allow superconductivity and magnetism to coexist. In 1952 Abrikosov published his findings in a Russian scientific journal.

Abrikosov then devoted his attention to examining the magnetic properties of Type-II superconductors. "The solution of the Ginzburg-Landau equation in the form of an infinitesimal superconducting layer in a normal sea of electrons was already contained in their paper," Abrikosov wrote in an article for *Physics Today* (January 1973). "Starting from this solution I found that below the limiting critical field, which is the stability limit of every superconducting nucleation, a new and very popular phase arose, with a periodic distribution of the [wave] function, magnetic field, and current. I called it the 'mixed state.'" Abrikosov devised mathematical equations that explained how a magnetic field successfully penetrated Type-II superconductors and was able to coexist with superconductivity. "By an insightful analysis of the Ginzburg-Landau equations he was able to show vortices may form in the spatial distribution of the order parameter and how a magnetic field through these can penetrate the superconductor," Professor

Mats Jonson said in his presentation speech to the Nobel laureates, as posted on the Nobel e-Museum website. "The vortices are essentially of the same type as those we can see form in the water when we empty a bath tub."In 1953 Abrikosov shared his theory with Landau. Although he was initially intrigued by Abrikosov's research, Landau strongly rejected the idea that vortices allowed magnetism to penetrate the superconductor. Abrikosov decided to postpone publishing his paper. "I put it in a drawer, but I did not put it in the wastepaper basket because I believed in it," he recalled to Jeremy Manier and James Janega for the *Chicago Tribune* (October 8, 2003). In 1957 Abrikosov finally published his paper in a Russian scientific journal. He also read the paper at a conference in Moscow that was attended by several British physicists. "Nobody understood a single word," Abrikosov recalled in his article for *Physics Today*. "This could be explained however, by the fact that I had a terrible cold with high [fever] and had hardly had any idea myself of what I talked about." The same year, Abrikosov's paper was translated into English and published in the *Journal of Physics and Chemistry of Solids*. Unfortunately, the translated article contained numerous errors in the equations and the text.

Abrikosov's work, however, was gradually vindicated, as more Type-II superconducting metals, which can carry more electricity than Type-I materials, were discovered during the 1960s. "With his theory of vortices, Abrikosov invented a whole new discipline," Brian Schwartz, the vice president for research at the Graduate Center of the City University of New York (CUNY), told Guy Gugliotta for the *Washington Post* (October 8, 2003). "If you read his [1952] paper today, you'll see it has everything. You're just amazed at how much he did."

### Significance

Superconductivity has led to numerous technological advancements, including magnetic resonance imaging (MRI), which provides medical doctors with high-resolution scans of the human body; particle-beam accelerators; and high-speed magnetic levitation trains.

Abrikosov was eventually named a senior scientist at the Institute of Physical Problems. In 1965 he became the head of the L. D.

Landau Institute of Theoretical Physics of the Academy of Sciences in Moscow. Abrikosov has also taught at Moscow State University, Gorky University, and the Moscow Physical English Institute. In 1988 Abrikosov became the head of the Institute of High Pressure Physics in Moscow. Disillusioned by life in the Soviet Union, Abrikosov came to the United States in the spring of 1991, months before the collapse of the USSR. He joined the Argonne National Laboratory in Illinois, as the Argonne Distinguished Scientist at the Condensed Matter Theory Group of the Materials Science Division. Explaining his decision to immigrate to the United States, Abrikosov told Margaret Shapiro for the *Washington Post* (November 23, 1991), "If you spend all day trying to get a car fixed and trying to find food, it doesn't stimulate theoretical research." At the Argonne National Laboratory, Abrikosov has pursued research in the fields of high-temperature superconductors and colossal magnetoresistance (CMR) manganates.

Arbikosov has published several books on physics and numerous scientific papers. He and his wife, Svetlana, are the parents of three children. Abrikosov lives in Lemont, Illinois.

## Further Reading

Abrikosov, A. A. "My Years with Landau." *Physics Today* Jan. 1973: 56–60. Print.

"And the Nobels go to … Illinois." *Chicago Tribune* 8 Oct. 2003: 1. Print.

Shapiro, Margaret "Ablest Soviets Flee for Better Lives; Economics, Not Dissidence, Fuels Accelerating Brain Drain." *Washington Post* 23 Nov. 1991: A1. Print.

# Boss, Alan P.

## American astrophysicist

**Born**: July 20, 1951; Lakewood, Ohio

Until the mid-1990s astronomers had no evidence for the existence of extrasolar planets—planets that orbit stars other than our sun—and thus could not know whether they are plentiful or scarce or exist at all. Thus, the only planetary system available to them for study was our own—the solar system. Then, in 1995, thanks to advances in technology, two Swiss astronomers, Michel Mayor and Didier Queloz, detected the presence of a planet orbiting the star 51 Pegasi; since then, more than seven hundred extrasolar planets have been discovered—not directly (except in an extremely few cases), but through measurable effects on their stars. The highly varied characteristics of the extrasolar planets detected to date have greatly surprised astronomers, astrophysicists, and others who study outer space; previously, they would have dismissed some of them as theoretically impossible. Many scientists now think it likely that Earth-like planets with some form of life exist elsewhere in the Milky Way (or in the hundreds of billions of other galaxies in the universe). Among them is the astrophysicist Alan P. Boss, a theoretician who has worked at the Carnegie Institution in Washington, DC, since 1982 and has served as an adviser on many National Air and Space Administration (NASA) panels. "From ground-based observations, we know that Earth-like planets are going to be quite common," Boss told Claudia Dreifus for the *New York Times* (July 20, 2009). "Estimates are that 'earths' probably occur [with] 10 to 20 percent of the stars. My feeling is that if you have that many earths and you have some prebiotic soup, comets that bring in the organic chemicals that you need to have life, something is going to grow. You might not always end up with dinosaurs and cavemen, but there are going to be planets out there that will have primitive life. Life on Earth is so vigorous and so able to thrive and fill every niche, how could it not be elsewhere? Give life a few billion years and, under the right conditions, something is going to happen, at the very least some sort of primitive bacteria like what we find in the geysers at Yellowstone [National Park, in Wyoming]."

Boss's theories about extrasolar planets are inextricably linked to another of his primary research interests: the events, starting some 4.6 billion years ago, that led to the formation of the planets of our solar system. The solar system consists of the sun and its eight major planets—Mercury, Venus, Earth, and Mars, which are nearest to the sun, and the more-distant Jupiter, Saturn, Uranus, and Neptune. (The solar system also contains a total of more than 140 moons, every planet except Mercury and Venus having at least one, and hundreds of thousands of comets, asteroids, and meteoroids. Pluto, discovered in 1930, was thought to be the ninth planet until 2006, when its status was reduced to that of a dwarf planet.) Theories about how the planets came to be must take into account the differences between the four planets nearest to the sun, known as the terrestrial planets, whose surfaces and topmost layers are solid and rocky, and the other four, the gas giants, whose surfaces and topmost layers are gaseous; they must also explain why the compositions of Uranus and Neptune differ significantly from those of Jupiter and Saturn. Boss believes that neither of the two theories of planet formation most commonly held among scientists—the core-accretion theory and the gas-accretion theory—adequately explains the formation of the solar system's gas giants or the extrasolar planets so far discovered, nearly all of which are gas giants. Boss is prominent among the few astrophysicists who have promoted the theory that the gas giants formed by a third process, dubbed disk instability. "Understanding how our solar system formed—that's my wild dream . . . ," Boss told Corey S. Powell for *Discover* (January 12, 2009). "It's taken thirty years, but I'm starting to think I understand what happened. I expected at some point I'd get old and not care anymore. But I care more and more. It's just amazing."

The University of Arizona astrophysicist Steven Kortenkamp told Keay Davidson for the *San Francisco Chronicle* (June 17, 2002) that the discoveries of extrasolar planets and the disk-instability theory "are making it very difficult to stick to the party line endorsing the so-called standard model" of planetary formation. "Alan Boss is one of the very few mavericks going out on a limb with a totally new theory . . . ," Kortenkamp declared. "Hopefully, we will know if Alan is right in a matter of a few years, maybe a decade." Harold F. Levison, at the Southwest Research Institute in Boulder, Colorado, who studies the behavior of planets and other objects in the solar system, told Robert Roy Britt for Space.com (July 9, 2002), "The thing we've learned in the last couple years is that the standard model [for the formation of planets] cannot work. I applaud Alan for trying to think in a new direction. I happen to think it's not right. But it's ideas like this that the science needs to evolve to the right answer." Boss is the author of *Looking for Earths: The Race to Find New Solar Systems* (1998) and *The Crowded Universe: The Search for Living Planets* (2009).

## Early Life and Education

The son of Paul Boss and Marguerite May (Gehringer) Boss, Alan Paul Boss was born on July 20, 1951, in Lakewood, Ohio, and raised in Florida. His sister, Barbara, he told *Current Biography*, "was just as much of a science nerd as I was, and with her being three years older, I followed in her footsteps in high school and college as an overachiever." After Boss completed all the science courses offered at his high school, he volunteered to assist a teacher in the chemistry lab. He was not interested in space then, he told *Current Biography*, because the "the bright lights, clouds, and night fogs" where he lived "mostly limited one's view of objects in the sky to the sun and the moon." After high school he enrolled at the University of South Florida, with the intention of majoring in oceanography. He abandoned that plan as a freshman, when he decided that the oceanography curriculum was boring. He studied physics instead, earning a bachelor's degree in that subject in 1973.

Boss then entered the University of California at Santa Barbara (UCSB), where he received a master's. degree in 1975 and a PhD in 1979, both in physics. He had intended to concentrate on high-energy physics, but during his second year there, he so impressed the astrophysicist Stanton J. Peale that Peale chose him to be his research assistant. Peale, whom Boss has described as a "celestial mechanician," "wanted me to work on the formation of the solar system, and that sounded like even more of a challenge than deciphering particle physics, so off I dove into the then uncharted waters of star and planet formation," Boss told *Current Biography*. He began by reading a book by the Soviet astronomer Viktor Safronov called *The Evolution of the Protoplanetary Cloud and Formation of the Earth and the Planets.* "Safronov's book simply started off with a young star and a rotating disk and then addressed the problem of planet formation," he told *Current Biography*. "I wanted to know how that star and disk formed

---

**"Life on Earth is so vigorous and so able to thrive and fill every niche, how could it not be elsewhere?"**

---

in the first place, and that meant studying how rotating clouds collapse to form not only single stars with disks, but also binary and multiple star systems." (By "disk," Safronov and Kirk were referring to the cloud of gas that orbits a developing star and gets flattened into a disk as the cloud rotates.)

## Life's Work

From 1973 to 1974, Boss was a teaching assistant, and for the next five years a research assistant, at UCSB; his title was changed to post-doctoral researcher after he earned his PhD. That year he left UCSB to join the division of space science at NASA's Ames Research Center in Moffett Field, California, where he worked as a resident research associate. He told *Current Biography* that his "main accomplishment" at Ames was the creation of a three-dimensional model of radiative transfer, the transport of electromagnetic radiation from one molecule or other object to another. In 1982, Boss took on the position of staff associate in the Department of Terrestrial Magnetism at the Carnegie Institution. He has worked there as a full-fledged member of the staff since 1983. Grants from NASA, since 1982, and from the National Science Foundation, since 1984, have funded his research, which has focused on the formation of planetary and stellar systems and extra-solar planets.

Astronomers believe that our star, the sun, and its planets started to form around 4.6 billion years ago. The process is believed to have begun when shock waves produced by a supernova—an exploding star—squeezed a nearby nebula, a gigantic cloud of gas (primarily hydrogen and helium) and dust, causing it to start to collapse. As the cloud collapsed, gravity pulled its contents closer together, and it began to spin (just as a spinning ice skater rotates faster as she pulls her arms closer to her body); as it spun, it flattened into a disk. Much of the matter in the disk was pulled into its center, where it became hot and formed the sun. Most of the remaining matter in the disk formed the eight planets of the solar system.

According to the core-accretion theory, the dust particles in the disk collided to form increasingly larger particles, growing to boulder size and then, as more matter stuck to them, to about six miles in diameter, when they became what are known as planetesimals; at that point their gravity had increased to the point that they experienced "runaway" growth. Their final sizes depended on their distance from the sun and their particular compositions and densities. According to the gas-accretion theory, as Boss explained to Corey S. Powell, "If there's gas around and the bodies get large enough, perhaps something on the order of ten Earth masses or so, then you can start pulling some gas in on top of your rocky core and make something that looks like a gas giant planet, like Jupiter. You just have to build that core fast enough to be able to pull in gas while the gas is still there." Boss believes that in the cases of some gas giants, the core cannot form rapidly enough to retain the surrounding gas. In the late 1980s, after working with computerized three-dimensional models, he offered another theory. The flattened disk, he proposed, had spiral arms, which, he told *Current Biography*, might have formed "self-gravitating clumps that could go on to form gas giant planets, rather than merely move gas and dust around the disk." Since then, he said, "I have been working to see if this serendipitous prediction might actually work in the real world. While disk instability is unlikely to explain the formation of all giant planets, the evidence is gathering that it forms at least a fraction of them."

## Disk-Instability Theory

The disk-instability theory has stirred a mixed response among scientists. Jack Lissauer, an astrophysicist at the Ames Research Center, told Keay Davidson that the "idea is hardly advanced enough to call it a model" and added, "Although Alan may believe that a paradigm shift has occurred, few others in the field agree with him." Steven Kortenkamp, on the other hand, told Davidson that Boss "is on to something interesting that is causing a much-needed shakeup in the field." In defense of his theory, Boss told Corey S. Powell, "Disk instability can form planets that look like Saturn and Jupiter just as well as core accretion can. So from that point of view, we now have two ways of making gas giant planets. And I think we need both of them. One or the other is not going to do it because there's such a wide range of systems." The discovery of the planet orbiting 51 Pegasi led Boss to conclude that yet another mechanism had been at work in its formation, because the planet, which is the size of Jupiter, is "too close to its sun to have formed there," he told Powell; it completely orbits the star in only 4.2 days. "One night I was lying in bed at 2 a.m. staring at the ceiling and it just sort of came to me," Boss recalled to Powell. "The damn thing must have migrated in and gotten left behind. There's just no way to understand its forming there."

In a review of Boss's first book, *Looking for Earths: The Race to Find New Solar Systems* (1998), Mira Dougherty-Johnson wrote for *Astronomy* (April 1999) that the author "possesses a unique insider's perspective that allows him to portray the reality of modern discovery. With the sympathy of a humble narrator, Boss takes readers on a thrilling journey of discovery. He details the actions of scientists who have created a new way of thinking about our solar system. And he shows how these remarkable finds have transformed this scientific field. . . . His book is full of humor, heartbreak, and a deep understanding of the ardor and luck that compose years of research. . . . As Boss looks upward and outward, he never loses sight of where he stands, grounded in humanity. As a result, the reader becomes not merely a receiver of Boss's vision, but a fellow explorer." David Morrison, who was then the director of astrobiology and space research at the Ames Research

Center, was similarly enthusiastic; in his review of *Looking for Earths* for *Natural History* (February 1999), he described Boss as "a gifted writer who gives us a detailed account of nearly a century of effort directed to distinguishing planets from stars, understanding how planetary systems form and evolve, and searching for the elusive evidence of extrasolar planets. . . . You will find no better introduction to one of the truly revolutionary developments in modern astronomy." In an assessment for *Science* (December 4, 1998), Jack J. Lissauer found fault with the book: among other flaws, Lissauer wrote, "Boss has not written a comprehensive review. Some of his terminology is quite unconventional. His presentation of the theory of planet formation is not balanced . . . and his interpretation of some observations differs from that of most observers and theorists in the field." Nevertheless, Lissauer praised the book for providing "a substantial amount of information about the search for extrasolar planets, as well as a good story."

In *The Crowded Universe: The Search for Living Planets* (2009), Boss discussed the probability that astronomers will find planets in the Milky Way hospitable to life. In a review for *Natural History* (May 2009), Laurence A. Marschall, an astronomer and physicist, wrote, "The tone of Boss's book . . . is excited and hopeful, but there's also a note of wry irony in his descriptions of the political trials astronomers have gone through trying to promote their research. And despite the successes of the past decade, Boss senses that it may be increasingly difficult for astronomers to attract the sums needed to continue the search for habitable planets. Readers of this book, I am certain, will hope his fears are unsubstantiated." The astrobiologist Lewis Dartnell wrote for *New Scientist* (March 14, 2009), "Boss recounts the exhilarating tale of the race to discover the first truly Earth-like exoplanet [extrasolar planet]. As *The Crowded Universe* unfolds, it brings alive the thrills and disappointments of bleeding-edge science, the fierce competition between American and European planet-hunting teams and the politics of billion-dollar research. Along the way we learn the latest theories on how planets form and just how astronomers detect distant worlds too faint to see. Frustratingly, the book is written as a series of chronologically dated sections which often lack cohesion. As a result, it feels as if Boss has simply lent us notes from his diary."

Boss is a member of the Science Working Group of NASA's Kepler Mission (named for the seventeenth-century German astronomer Johannes Kepler), whose spacecraft, launched in March 2009, is searching for Earth-like exoplanets among approximately 150,000 stars in the constellation Cygnus in the Milky Way. In particular, Boss is helping to "interpret [the craft's] discoveries in terms of the theory of planet formation," he told *Current Biography*. "The most certain discovery in the next five years is the determination of the fraction of Earth-like planets in our galaxy," he said.

Boss is a member of the American Academy of Arts and Sciences and a fellow of the American Association for the Advancement of Science, the Meteoritical Society, and the American Geophysical Union. He was the founding chairman of the International Astronomical Union's (IAU) Working Group on Extra-Solar Planets (WGESP) and is currently the president of WGESP's successor, known as IAU Commission 53 on Extrasolar Planets. Boss has been married to the former Catherine Ann Starkie since 1979. His wife is an accountant and has her own practice. They have two children, Margaret and Nicholas, and live in a suburb of Washington, DC. Of the 400,000 known asteroids, one—minor planet number 29137—bears the name Alanboss.

## Further Reading

Davidson, Keay. "Heated debate on formation of solar system/Theorist believes 9 planets created out of chaos, not calm." *San Francisco Chronicle* 17 June 2002: A6. Print.

Dreifus, Claudia. "Searching for Extraterrestrial Life." *New York Times* 21 July 2009: D2. Print.

Pawlowski, A. "Galaxy may be full of 'Earths,' alien life." CNN.com. Turner Broadcasting System, 25 Feb. 2009. Web. 5 Apr. 2012.

Powell, Corey S. "The Man Who Made Stars and Planets." *Discover Magazine* 12 Jan. 2009. Print.

# Chu, Steven

American physicist and secretary of energy

**Born**: February 28, 1948; St. Louis, Missouri

Steven Chu is considered one of the world's leading experts on climate change and the development of alternative, environmentally friendly sources of energy. He first garnered attention in 1997, when he was jointly awarded the Nobel Prize in Physics for his work in developing methods to cool atoms using laser light. In December 2008, US president-elect Barack Obama nominated Chu for the post of secretary of energy. Confirmed by the US Senate on January 20, 2009, he is the first person of Chinese descent to occupy the post.

## Early Life and Education

The second of three sons, Chu was born on February 28, 1948, in St. Louis, Missouri. He came from a family that valued learning. "Education in my family was not merely emphasized, it was our raison d'être," Chu wrote in an autobiographical essay posted on NobelPrize. org. "Virtually all of our aunts and uncles had PhDs in science or engineering, and it was taken for granted that the next generation of Chus were to follow the family tradition." His father, Ju Chin Chu, earned a chemical-engineering degree from the Massachusetts Institute of Technology (MIT) in Cambridge, where his mother, Ching Chen Li, studied economics. His maternal grandfather had also received an advanced degree in civil engineering, from Cornell University in Ithaca, New York. (Chu's parents had originally planned to return to China upon finishing college but decided to settle in the United States, due to the political unrest in their native country.) His older brother, Gilbert Chu, is a professor at the Stanford University School of Medicine; his brother Morgan Chu is a nationally known trial lawyer.

By 1950 Chu and his family had moved to Garden City on Long Island, New York; his father taught nearby, in the New York City borough of Brooklyn, at what is now the Polytechnic Institute of New

York University. From an early age Chu enjoyed constructing things with his hands. "I would be given for Christmas a model set of airplanes or boats and things, and I loved to put them together. I would ask my parents for things like Erector Sets," he said in an interview posted on the University of California, Berkeley website. "In many respects, my brothers and I are very similar, but in that respect, I seemed to love mechanical things in a way that was certainly nurtured by my parents, in that they said, 'Okay, he wants to do these things. We'll buy toys like that for him.' But my other two brothers didn't seem to like that."

Chu first developed an interest in science during his senior year at Garden City High School, when he took two advanced-placement courses, in calculus and physics. As a result of his A-minus average—well below family expectations—Chu, whose older brother was attending Princeton University and whose two cousins had been accepted to Harvard University, was denied admission to several Ivy League universities and enrolled instead at the University of Rochester, earning a BA degree in mathematics and a BS degree in physics in 1970. He then attended graduate school at the University of California (UC), Berkeley, with the intention of becoming a theoretical physicist, but he developed an interest in experimental physics after conducting a series of laser experiments under the supervision of Eugene Commins, his mentor.

After receiving his PhD in physics in 1976, Chu remained at UC Berkeley as a postdoctoral fellow until 1978. That year he joined the technical staff at AT&T Bell Laboratories in Holmdel, New Jersey, where he met his future collaborator Art Ashkin, who was credited with developing the world's first methods of trapping atoms with laser light. At that time Ashkin and a group of fellow Bell scientists conducted several experiments, in which they explored how to cool an object by directing a laser beam at it.

## Life's Work

Five years later, after funding for Ashkin's experiments was cut, Chu was promoted to head of the lab's quantum-electronics research department and collaborated with Ashkin on a series of experiments that involved manipulating atoms at low temperatures. "The [conventional wisdom] was first, you hold onto an atom; then you get it cold; and then you can do what you want with it. I said, 'Well, what if you reversed it? What if you cooled down the atom first? Don't hold onto it, but maybe in the process of cooling it down, it's going to hang around for enough time that you can have a chance of grabbing onto it,'" he said during the interview posted on the UC Berkeley website. "And so [after] a little calculation I said, 'Holy smokes. This looks like it's going to work.'"

In 1987, after nine years at Bell Labs, Chu accepted a position as a physics professor at Stanford University in California, where he continued his research in low-temperature physics. The laser techniques he developed to cool and trap atoms and molecules earned him the 1997 Nobel Prize in Physics, which he shared with Claude Cohen-Tannoudji and William D. Phillips. While at Stanford, Chu also helped establish the school's Bio-X program, which assembles scientists from the physics, chemistry, biology, and engineering fields. Chu also served as chair of the Physics Department (1990–93 and 1999–2001).

In 2004 Chu was appointed the director of the Lawrence Berkeley National Laboratory, a US Department of Energy national lab located at UC Berkeley. Under Chu, the lab became an important center of research into alternative energies, energy efficiency, biofuels, and climate change. An advocate of reducing greenhouse gases, Chu has warned about the dangers of climate change and global warming, which he predicts will lead, if unchecked, to water and food shortages, resource wars, and chronic displacement of the world's poorest people. (The rise in temperatures due to climate change can lead to variable rainfall patterns. In developing countries it has led to decreased rainfall levels, reducing the water supply and making it difficult for crops to grow. In other areas it has resulted in rising sea levels and increased flooding, which can destroy the quality of water and contribute to waterborne diseases.) With the aim of avoiding such outcomes, he established the Helios Project, a new facility dedicated to researching low-cost solar and other sustainable, alternative, and renewable energy solutions, including advanced biofuels and artificial photosynthesis. Listing the facility's accomplishments, Steven Mufson wrote for the *Washington*

*Post* (December 12, 2008), "The laboratory's scientists, including eleven Nobel laureates, have altered yeast and bacteria into organisms that produce gasoline and diesel, improved techniques for converting

> **"Education in my family was not merely emphasized, it was our raison d'être."**

switchgrass into the sugars needed to produce transportation fuel, and used nanotechnology to improve the efficiency of photovoltaic cells used in solar panels, among other projects." Additionally, Chu is the founder of the Energy Biosciences Institute.

On December 11, 2008, president-elect Barack Obama nominated Chu to be the next US secretary of energy; he was confirmed on January 20, 2009, and sworn in on the following day. In that post, he is responsible for implementing Obama's ambitious energy policy, which includes plans to invest $150 million over the next ten years to create five million new jobs and to produce one million electric cars by 2015, among other projects. Chu has been called "America's first clean-energy secretary," and his mission, as Michael Grunwald described it for *Time* (August 23, 2009), is "part green evangelism, part venture capitalism, and part politics." Both scientists and environmentalists expressed enthusiasm for his appointment, especially because past energy secretaries have often been, in Grunwald's words, "political loyalists with little energy expertise." Obama and Chu have listed several priorities, including promoting and institutionalizing energy efficiency; investing in wind, solar, and other renewable forms of energy; and providing incentives for privately funded research and development aimed at ending the nation's dependence on fossil fuels and making meaningful progress in the struggle to curtail global warming and climate change.

Chu has admitted to being politically naive and has expressed frustration at the obstacles seemingly inherent in the day-to-day functioning of the bureaucracy in Washington. Some environmentalists have expressed the fear that clean-energy legislation will take a back seat to economic problems and the need to reform the nation's health-care system. Nevertheless, both the president and Chu have argued that the production of clean energy and measures to reverse global warming can create jobs and wealth.

A member of the Copenhagen Climate Council and a former adviser to the directors of the National Institutes of Health and the National Nuclear Security Agency, Chu has written or cowritten more than two hundred scientific papers and holds ten honorary degrees. From his first marriage, which ended in divorce, he has two sons, Michael and Geoffrey. From his marriage to the physicist Jean (Fetter) Chu, in 1997, he has a stepson and a stepdaughter.

## Further Reading

Goodell, Jeff. "The Secretary of Saving the Planet." *Rolling Stone* 25 June 2009: 58+87. Print.

Kintisch, Eli. "Nobelist Gets Energy Portfolio, Raising Hopes and Expectations." *Science* 19 Dec. 2008: 1774–75. Print.

May, Mike. "Interview: Steven Chu." *American Scientist* Jan./Feb. 1998: 22. Print.

Mufson, Steven. "Concern for Climate Change Defines Energy Dept. Nominee." *Washington Post.* The Washington Post Company, 12 Dec. 2008. Web. 5 Apr. 2012.

Solomon, Deborah. "The Science Guy." *New York Times Magazine* 19 Apr. 2009: MM14. Print.

Wald, Matthew L. "Steven Chu." *New York Times.* The New York Times Company, 5 Dec. 2008. Web. 5 Apr. 2012.

# Cousteau, Fabien

## Aquatic filmmaker and oceanographer

**Born**: October 1967; Paris, France

For over half of a century, the name Cousteau has been synonymous with the exploration of the undersea world, so it comes as no shock that Fabien Cousteau, as a third-generation oceanographic explorer and documentary filmmaker, has chosen the same career path of his esteemed father Jean-Michel Cousteau and legendary grandfather Jacques-Yves Cousteau. Cousteau, a graduate of Boston University (BU), has managed to carve out his own unique identity within the Cousteau family: he has made two nationally aired documentaries, *Attacks of the Mystery Shark* (2002), about the real-life Jersey Shore shark attacks of 1916, and *Shark: Mind of a Demon* (2006), in which he observed the behavior of sharks from inside a shark-shaped submarine. Cousteau has also been actively involved in numerous marine conservation organizations and programs in efforts to protect the world's oceans and endangered marine species. He partnered with his father and sister Celine for the multipart Public Broadcasting Service (PBS) series *Ocean Adventures*, which aired from 2006 to 2009. Cousteau has also launched other television and media projects through his production company Natural Entertainment. In 2010 he created a nonprofit program called Plant-A-Fish, which leads initiatives to repair marine environments.

Though some observers would view the task of living up to a legacy like that of Jacques-Yves Cousteau's to be more of a burden than a blessing, Fabien Cousteau has maintained that he has never tried to fill his grandfather's shoes and was never pressured into going into the family business. Instead, he has chosen to carry on the family legacy out of passion and as a matter of personal responsibility. He told journalist Ron Gluckman, in an article for the luxury travel magazine *Centurion* (November 2012) posted on Gluckman's official website, "I think once you take a peek under the great blue, it's hard to turn your back. I'm addicted to ocean exploration. We're facing some very real challenges in this world, and we have to start fixing things."

## Early Life and the Family Business

The older of the two children of Jean-Michel and Anne-Marie Cousteau, Fabien Cousteau was born in October 1967 in Paris, France, the first grandchild of the pioneering ocean explorer, filmmaker, and researcher Jacques-Yves Cousteau. Considered the founding father of scuba diving, Jacques-Yves Cousteau coinvented the Aqua-Lung, an underwater breathing apparatus now known as the diving regulator, with the French engineer Emile Gagnan in 1943. He was also instrumental in the development of the first underwater cameras. He won Academy Awards for his groundbreaking undersea documentaries *The Silent World* (1956, codirected with the noted French filmmaker Louis Malle) and *World Without Sun* (1964), and rose to international fame as the host and star of the cult television series *The Undersea World of Jacques Cousteau*, which aired from 1968 to 1976 and inspired generations of ocean explorers. A prolific documentary filmmaker, he produced around 140 documentaries before dying, in 1997, at the age of eighty-seven.

Jacques-Yves Cousteau's eldest two sons, Jean-Michel and Philippe, both followed in his footsteps. Jean-Michel, Fabien's father, served for over ten years as executive vice president of the Cousteau Society, his father's Hampton, Virginia–based nonprofit environmental education organization, before venturing out on his own as a documentary film producer and marine conservationist. He established the underwater production company Deep Ocean Odyssey and is the founder and president of the Ocean Futures Society, a nonprofit organization dedicated to educating people about the world's oceans. Philippe Cousteau, meanwhile, was an ocean explorer and environmentally-conscious documentary filmmaker who coproduced and starred in many of his father's films before dying tragically in a PBY Catalina flying boat crash in the Tagus river near Lisbon, Portugal, in 1979. He had two children, Alexandra (born in 1976) and Philippe Jr. (born in 1980, six months after his father's death), both of whom have worked to carry on his legacy as oceanographers and environmentalists through their organization EarthEcho International. In September 2006 Philippe Jr. was working with the famed Australian "Crocodile Hunter" Steve Irwin as an apprentice cohost on the iconic television personality's nature documentary *Ocean's Deadliest* when Irwin was fatally pierced in the chest by a stingray while snorkeling off the coast of Queensland, Australia. Philippe went on to finish hosting the documentary himself and later served as chief ocean correspondent for the nature-centric cable channels Animal Planet and Planet Green.

Along with his younger sister, Celine, Fabien Cousteau grew up immersed in the family business. He spent a good portion of his childhood observing his grandfather on expeditions aboard his famous research ship *Calypso*, where both his parents worked—his father as an explorer and researcher, and his mother as an expedition photographer. Meanwhile, he started diving "before many kids take off their training wheels," as Tim Stoddard wrote for *Bostonia* (Summer 2003), the quarterly magazine for alumni of Boston University. When Cousteau was four years old, his grandfather built him custom scuba gear, and he began diving in his family's swimming pool in Sherman Oaks, California; soon afterward he was diving in the ocean. At seven, Cousteau went on his first *Calypso* expedition with his father and grandfather to Papua New Guinea. His first job was scrubbing barnacles off his grandfather's ship, but on later expeditions, he would be assigned to less menial, even glamorous roles as part of the crew. Cousteau recalled, as quoted in the *Dublin, Ireland, Sunday Independent* (November 9, 2008), "While other kids were at Disneyland, I was learning how to steer the *Calypso*."

While accompanying his grandfather and father on expeditions to some of the world's most remote places, Cousteau was instilled with an appreciation and respect for the ocean and its ecosystems. From an early age he developed a particular fascination with sharks, and he fondly remembers wanting to actually become a shark as a child after reading a book about the Belgian comic book character Tintin and his dog Snowy, titled *Red Rackham's Treasure* (1944), in which the two embark on a treasure hunt in a shark-shaped submarine; the book would later serve as the inspiration for Cousteau's 2006 documentary *Shark: Mind of a Demon*. His fascination was piqued even further after watching director Steven Spielberg's 1975 blockbuster thriller *Jaws* for the first time, when he snuck into a viewing of the film on a cruise ship where his father was giving a lecture. Based on author Peter Benchley's 1974 best-selling novel of the same name, the film helped perpetuate popular misperceptions about sharks and their behavior and was also largely responsible for scaring away generations of people from the water. Cousteau snuck into the film against his parents' orders, for which he was consequently grounded, he recalled to Stoddard, adding, "I walked out of there completely mystified because it went against everything I'd ever been taught." The film would ultimately inspire him to become a tireless advocate for sharks.

## Education

Cousteau has noted that he moved twenty-four times growing up, due to his family's business. Consequently, he attended schools all over France and the United States. "Making new friends every time I started a new school may have been challenging," he wrote in an article for the *London Telegraph* (August 12, 2011), "but I got to experience the weird things out in the ocean world that most children never had a chance to." By the age of twelve, Cousteau had become a regular crew member on the *Calypso*, which resulted in him spending the rest of his teenage years in and out of school.

In 1987, after attending an all-male boarding school in Connecticut, Cousteau enrolled at Boston University's College of General Studies (CGS), in Boston, Massachusetts, which offers a two-year general education program. It was during this time that he first started exploring career endeavors outside of the family business, encouraged by his grandfather and parents. "My life experience up until then had been nothing but marine biology," he recalled to Stoddard. "So I wanted a flavor of something else. BU was a turning point in my life, because those were the years when I established myself outside of my family." After completing his core curriculum requirements at CGS in 1989, he entered BU's Metropolitan College (MET), where he majored in environmental economics. He earned a bachelor's degree in environmental economics from MET in 1991.

## Career

After graduating from BU, Cousteau spent some time in fields outside the family business. He worked briefly in graphic design and sales management before joining the Burlington, Vermont–based environmental products development company Seventh Generation, where

> **"I think once you take a peek under the great blue, it's hard to turn your back. I'm addicted to ocean exploration."**

he worked in new product development. In that division he helped develop and market environmentally sustainable products. Cousteau has said that he purposely took on jobs after college that would force him to be more assertive and that would help him later on in his dealings as an independent filmmaker and businessman. He told Stoddard, "When you're sitting at the negotiating table with ABC executives, or raising funds for an expedition, it's essential to be able to speak their language, and to be able to translate your passion into words they understand."

Cousteau enjoyed success in the business world, but the work he was doing ultimately made him feel "kind of empty," as he told Ann Oldenburg for *USA Today* (June 27, 2006). As a result he returned to the family business with a reinvigorated sense of passion and commitment. In the late 1990s he joined Deep Ocean Odyssey, the family's production company, and began working on documentaries with his father. He first came to international attention as an ocean explorer in 2000, when he accompanied his father on a documentary shoot in South Africa. Two years later he was pegged to host his first television special, *Attacks of the Mystery Shark*, as a correspondent for MSNBC's *National Geographic Explorer* documentary series. The hour-long special examined the deadly shark attacks that had occurred along the coast of New Jersey during a two week span in July 1916. The attacks had claimed four lives and wounded another, with three of those victims having been attacked upstream in Matawan Creek, located several miles from the ocean. The attacks were blamed on a great white shark and created a nationwide frenzy as people began perpetuating the theory that all sharks were human-eating predators. These attacks served as the inspiration for Benchley's novel *Jaws* (1974) and Steven Spielberg's 1975 film of the same name, as well as for numerous other media representations of sharks.

For *Attacks of the Mystery Shark*, Cousteau traveled to the site of the attacks in New Jersey, as well as to India's Ganges River Delta, the Bahamas, and Pacific Ocean, in efforts to give viewers a better understanding of sharks and their behavior. Contrary to the popular belief that a great white shark had been responsible for the attacks, Cousteau's film suggests that they were most likely committed by a rogue bull shark, a smaller but more aggressive and unpredictable shark species. The film also contends that shark attacks are almost always a case of mistaken identity, with sharks mistaking humans for fish. Viewers learn that of the eighty or so shark attacks against humans each year, only a handful are fatal, and that the public's grave misperceptions about sharks is contributing to their demise. *Attacks of the Mystery Shark* attracted mainstream media attention and Cousteau's good looks led *People* magazine to name him "Sexiest Man of the Sea" in 2002.

Cousteau spent several years working on his next project, a self-produced documentary centering on the behavior of great white sharks. For the project, he revisited the premise of his favorite Tintin story to create a first-of-its-kind shark-shaped submarine. In order to realize his vision, Cousteau enlisted the help of family friend Eddie Paul, a renowned Hollywood engineer, inventor, and prop maker who has coordinated stunts and special effects for numerous feature films, including *Grease* (1978), *E.T.: The Extra-Terrestrial* (1982), and *The Terminator* (1984). (Paul had previously built a robotic shark for Cousteau's father in the 1980s, but it was destroyed by a great white shark off the coast of Southern Australia when it started malfunctioning.) The end result was a fourteen-foot, 1,200-pound wet sub, or a submarine designed to flood when underwater, that looked, moved, and sounded as much like a great white shark as possible. The sub featured hidden video cameras and a high-pressure pneumatic propulsion system, and it was covered in a rubber-like skin called Skinflex. The prototype, named "Troy" and costing over $100,000, allowed Cousteau to swim silently with great white sharks and observe them in their undisturbed natural habitat, as opposed to filming them from a conventional steel cage. Because Troy's steel-enforced cavity was built to fill with water after being submerged, Cousteau had to wear a wet suit and use scuba gear in order to breathe while piloting the mechanism.

The resulting one-hour special *Shark: Mind of a Demon* premiered on CBS in the summer of 2006 and helped debunk the popular myth that great white sharks are mindless killers and eating machines. It chronicled the trials and tribulations of Cousteau and his crew, including Paul, the shark expert Mark Marks, and Cousteau's sister Celine, following them over a two-year period as they tried to launch Troy in the great white shark–infested waters and feeding grounds near Guadalupe Island off the west coast of Mexico in the Pacific Ocean. Despite experiencing several mechanical issues with Troy during the course of filming, Cousteau was able to make some successful dives in the shark sub without being harmed.

The year 2006 also saw Cousteau collaborate with both his father and sister for a three-year PBS documentary series called *Ocean Adventures*, which followed the trio and their expedition team as they examined different types of marine life from all over the world. The series premiered in April of that year with the special *Voyage to Kure*, which explored the remote Kure Atoll of the northwestern Hawaiian Islands. Other specials in the *Ocean Adventures* series include *Sharks at Risk* (2006), which focused on gray sharks in French Polynesia and great white sharks in South Africa; *The Gray Whale Obstacle Course* (2006), which followed gray whale migration routes; *America's Underwater Treasures* (2006), a two-part episode that investigated all thirteen of the US National Marine Sanctuaries and the Northwestern Hawaiian Islands Marine National Monument; *Return to the Amazon* (2008), a ten-month expedition through the Amazon that explored the devastating effects of deforestation and development in the biodiverse region; *Sea Ghosts* (2009), an examination of the effects of climate change on Arctic-dwelling Beluga whales; and *Call of the Killer Whale* (2009), which shed light on the similarities between humans and orcas, or killer whales.

Cousteau has continued to lead and take part in expeditions all over the world. He has also cofounded a production company called Natural Entertainment, which develops television and other media projects related to ocean exploration and environmental awareness. In 2010, in honor of the one-hundredth anniversary of his grandfather's birth, he established Plant-A-Fish, a nonprofit organization that works to educate and empower individuals and communities through the replanting of ocean animals and plants. Cousteau explained to Gluckman that he wants the organization to "help influence the world to be a better place, specifically, to reconnect with the ocean and to feel the same passion most people felt when they first saw my grandfather's documentaries. It's a daunting task, but we can do it." Plant-A-Fish has launched programs to plant one billion oysters in the New York City area, one billion sea turtles in El Salvador, and one million corals in Florida and the Maldives.

Cousteau's other endeavors include writing and public speaking. He has written articles for the *Huffington Post* and other publications and is working on a children's book trilogy "about the future of our planet, a fictional account based on reality," as he told Gerri Miller for Mother Nature Network. He has also spoken on a wide range of environmental issues at conferences and educational institutions all over the world. While acknowledging his grandfather's lasting influence on his life, Cousteau told Stoddard, "I don't want people to think of me as Jacques Cousteau's grandson. I'd like to continue in the spirit of my grandfather, but no one can fill his shoes, and I wouldn't pretend to do that." He added, in his essay for the *London Telegraph*, "I remember clearly my grandfather saying to me when I was a child that the two fundamental things we need to live are air and water, and we are treating both as a garbage can. He created a consciousness about our oceans, and taking better care is more urgent now than ever before, so I try my best to carry on his message."

## Personal Life

Cousteau divides his time between New York City and the South of France. He serves on the board of the New York Harbor School, a public high school located on Governor's Island that offers special marine-based programs, and he is a member of the Water Innovations Alliance, which works to raise awareness of water issues by bringing them to the attention of corporate executives. Cousteau spends most of the year traveling but enjoys mountain biking and wind surfing in his spare time. He is also an avid rider and collector of vintage motorcycles, with at least eight vintage bikes in his collection. He told Ingrid Skjong for *Gotham Magazine* (August 1, 2009) that he goes to bed "exhausted" every night and wakes up "every morning ready to do it again."

## Further Reading

Cousteau, Fabien. "Biography." *Fabien Cousteau: The Adventure of Discovery*. Fabien Cousteau, n.d. Web. 5 Apr. 2012.

Cutter, Kimberly. "Who Is the Nouveau Cousteau?" *Men's Journal* (2008): n. pag. Print.

Finn, Robin. "PUBLIC LIVES; Heir to an Undersea World, Swimming With Sharks." *New York Times* 31 July 2002: 2. Print.

Miller, Gerri. "Cousteau, the Next Generation." *Arts & Culture*. MNN Holdings, LLC, 21 Apr. 2009. Web. 5 Apr. 2012.

Stoddard, Tim. "The Underseen World of Fabien Cousteau." *Bostonia*. Boston University, Summer 2003. Web. 5 Apr. 2012.

# Davies, Paul

## English physicist

**Born**: April 22, 1946; London, England

"Most people think that as science advances, religion retreats," the professor and physicist Paul Davies once said, as quoted by Gustav Niebuhr for the *Houston Chronicle* (March 9, 1995). "But the more we discover about the world, the more we find there's a purpose or design behind it all." For years Davies has argued that the more we learn of the origins of the universe through science, the more we will understand God. "Science is about explaining the world, and religion is about interpreting it," he told Susan Long for the *Singapore Straits Times* (October 16, 1998). "There shouldn't be any conflict." An internationally respected authority and bestselling popular-science author, Davies is concerned with humanity's relevance in the universe, the origins of existence, life on other planets, and the physical possibility of time travel. He has also made important contributions to theories about the development of the universe.

Davies is a member of what he describes as an alternative academy of scientists, a loose consortium of professionals whose work has helped to construct, as he told Harriet Swain for the *Times Higher Education Supplement* (August 7, 1998), a "view of the world which is not religious in the conventional sense—but is more comforting than the bleak reductionism of most science over the past 300 years." Other members of this alternative academy include the scientist James Lovelock and the Oxford professor of mathematics Roger Penrose.

Davies is the author of over twenty books, many of which have been translated into dozens of languages and have become international bestsellers. His books seek to demystify the laws of physics and the universe by simplifying advanced scientific theory and examining it in philosophical terms. He has published over one hundred research papers on such topics as cosmology, gravitation, and quantum field theory. He is also much in demand for speaking engagements and has lectured throughout Asia, North America, Europe, and Australia, as well as on public television in England and Australia.

Davies hopes that by popularizing the concepts behind physics, the mysteries of science will be shed to reveal evidence of a "grand design," he told Bill Broadway for the *Washington Post* (March 11, 1995). He seeks to repudiate the notion that the universe is "nothing but a gigantic collection of stupid particles colliding like cogs in a machine and that human beings are locked in this cosmic juggernaut."

## Early Life and Education

Paul Charles William Davies was born in London's Charing Cross Hospital on April 22, 1946, and raised in the Hampstead Garden and Finchley sections of London. His father, Hugh, was a clerk for the London-area Saint Pancras Cemetery and later for the Camden Housing Department. His mother, Pearl, was a shorthand typist. By age fourteen, as a student at Woodhouse Grammar School in London, Davies knew he wanted to pursue a career in theoretical physics and cosmology, and asked his chemistry teacher how he could go about becoming an astronomer. The teacher was not very encouraging, telling the young Davies that he did not know anyone in the field and did not think the profession paid particularly well. Davies did not receive much support for his scientific ambitions from family members, either. "There were no scientists in my family, no academics even," he told Susan Long. "I was the first person to go to university in my family. . . . I used to drive [my parents] crazy with my why, why, why questions. My mother would say: 'Because God made it that way and that's that.' But I was never happy with that answer."

Davies was raised loosely in the Anglican faith—"My parents went to church twice a year, at Christmas and maybe Easter," he told Susan Long—but he gave up religion in his teens while grappling with questions about the existence of God and pondering the meaning of human existence. He told Harriet Swain, "As a teenager, I used to lie in bed wondering where do I come from, what will happen when I die and so on. I used to worry a lot about free will. Anything that was hidden interested me." Davies found answers to some of his questions in the work of the cosmologist Fred Hoyle, whose book *Frontiers of*

*Astronomy* he read at sixteen. "Hoyle presented science as a fast-paced detective story with sweeping implications," Davies said, as quoted in the Melbourne, *Australia Sunday Age* (December 16, 2001). "The passion and excitement of being at the forefront of research shone from every page. I was already wondering about making a life in science, and this book confirmed it for me." Now an "avowed theist," as characterized by Bill Broadway, Davies considers himself one of many individuals "who never go into a church or a synagogue or a mosque but nevertheless are religious in the broadest sense."

## Life's Work

After graduating with honors in physics from University College London in 1968, Davies went on to earn his PhD in theoretical physics from the same institution in 1970. Soon after, he joined the Cambridge Institute of Theoretical Astronomy, where he worked with the renowned physicist Stephen Hawking on the thermodynamics of black holes and the big bang theory. Around this time he wrote articles on the nature of time for the *Physics Bulletin*; *New Scientist* and *Nature* asked him for regular contributions, the latter for a "news and views" column. In 1972 Davies joined King's College London as a lecturer in applied mathematics, a post he held until his appointment as professor of theoretical physics at the University of Newcastle upon Tyne, in 1980.

In 1983 Simon and Schuster published Davies's *God and the New Physics*, which discussed the impact of science on religious issues. In a review for the *New York Times* (December 9, 1984), Timothy Ferris wrote, "Mr. Davies knows the arcana of physics the way a plumber knows wrenches, and he can make sense out of quite daunting ideas. He proves to be one of the most adept science writers on either side of the Atlantic."

In 1988 Davies published *The Cosmic Blueprint: New Discoveries in Nature's Creative Ability to Order the Universe*, in which he wrote, as quoted by Lee Dembart for the *Los Angeles Times* (March 18, 1988), "There must be new general principles—organizing principles over and above the known laws of physics—which have yet to be discovered. We seem to be on the verge of discovering not only wholly new laws of nature, but ways of thinking about nature that depart radically from traditional science." Davies proposed in *The Cosmic Blueprint* that existence is not the result of accident and chance; instead, it is the product of a self-organizing universe. "Complex structures in biology are unlikely to have come about as a result of purely random accidents," he writes, as quoted by Lee Dembart. "Even the simplest living things are stupendously complex. The replicative machinery of

---

**"Science is about explaining the world, and religion is about interpreting it."**

---

life is based on the DNA molecule, which is itself as structurally complicated and intricately arranged as an automobile assembly line. If replication requires such a high threshold of complexity in the first place, how can any replicative system have arisen spontaneously?" Also in 1988, Davies published *Fireball*, his first and only foray into science fiction writing—"the kind of novel," John Christie wrote for the *London Guardian* (February 5, 1988), that "theoretical physicists occasionally seem compelled to write."

In 1990, angered by the British government's budgetary cuts in university-level science education, Davies accepted a professorship of mathematical physics at the University of Adelaide and moved his family to Australia. "There is no doubt whatever Australia is a far more cultured country than Britain," Davies told Peter Lennon for the *London Guardian* (April 6, 1990). "Science is seen as a proper cultural activity there. People recognize the value of pure research; they see that science is a cultural pursuit just as much as art or literature. [In England] it is no longer recognized as part of culture; it is simply seen as an adjunct to industry." Davies added, "[Australia] is a perfect country for children to grow up in. The educational system is just incomparably better. It has a clean environment; an absence of a class system and a combination of relaxed life style and a positive, optimistic view." Davies's work was immediately recognized by Australia's science establishment. In 1991 he was awarded the Australian Museum Australian Broadcasting Corporation Eureka Prize, bestowed on those whose work has helped to bring science to the broader public. In 1993 the University of Adelaide created a post specifically designed to suit Davies's field of research and expertise: he became the university's first professor of natural philosophy, a term once used to describe physics.

In 1992 Simon and Schuster published Davies's *The Mind of God: A Scientific Basis for a Rational World*, in which he revisited issues raised in *God and the New Physics* and *The Cosmic Blueprint*; in particular, *The Mind of God* focused on the nature of human consciousness and the beginnings of the universe. "In his quest for an ultimate explanation," stated the book's back-cover blurb, "Davies reexamines the great questions that have preoccupied humankind for millennia, and in the process explores, among other topics, the origin and evolution of the cosmos, the nature of life and consciousness, and the claim that our universe is a kind of gigantic computer. . . . His startling conclusion is that the universe is 'no minor byproduct of mindless, purposeless forces. We are truly meant to be here.' By the means of science, we can truly see into the mind of God." Marcia Bartusiak, in the *New York Times Book Review* (February 23, 1992), called *The Mind of God* "both stimulating and enlightening." It was awarded the 1992 University of New South Wales Press Eureka Prize and the 1995 Templeton Center for Theology and the Natural Sciences (CTNS) book prize. (Located in Berkeley, California, CTNS is a nonprofit international organization committed to the research and teaching of contemporary physics in relation to cosmology, technology, environmental studies, evolutionary and molecular biology, and ethics.)

In 1994 Davies, in his continuing venture to convey the excitement of the ever-changing universe to the general public, published *The Last Three Minutes: Conjectures about the Ultimate Fate of the Universe*, which presented various scenarios for the end of the universe—including cosmic implosion, entropic heat-death, and massive gravitational collapse in a final "Big Crunch." In his next book, *About Time: Einstein's Unfinished Revolution* (1995), Davies discussed the big bang and chaos theories as well as Einstein's principles of time.

For his contributions to religious thought through the study of science, Davies was awarded the Templeton Prize in May of 1995. The

Templeton Prize was created in 1973 by the American-born investor Sir John Marks Templeton, who sought to acknowledge individuals whose work helps demystify concepts surrounding spirituality, God, and the universe. The award, worth approximately $1,000,000, was presented by Prince Philip at Buckingham Palace. The judges presiding that year included the former US president George H. W. Bush and the former British prime minister Margaret Thatcher. This was the second of two encounters Davies has had with the former prime minister. He recalled the story of their first meeting in his prize address, as transcribed in *First Things* (August/September 1995): "I was fascinated to learn that one of the judges for the Templeton Prize was Baroness Thatcher. This is in fact the second time she has been involved in giving me a prize. The first occasion was in 1962, at the Speech Day of Woodhouse Grammar School in North Finchley, when she presented me with a copy of Norton's Star Atlas for doing well in my O level exams. I doubt if her Ladyship recalls the encounter, but I can trace my own decision to become a scientist more or less to that event." The judges, as quoted by Bill Broadway, called Davies "one of the world's most brilliant scientists. He works at the forefront of research in fundamental physics and cosmology. . . . He has initiated a new dialogue between science and religion that is having worldwide repercussions." (Previous winners of the Templeton Prize have included Mother Teresa and the Reverend Billy Graham.)

In 1996 Davies retired from his professorship at Adelaide to focus on his writing and speaking engagements. (He was in increasingly high demand for lectures throughout Europe, Australia, and the United States.) "I have tried to straddle the world of academia and the real world," Davies told Peter Hackett for the *Advertiser* (June 10, 1997), but academic life, he admitted, had become "somewhat claustrophobic" and he was finding it a task to maintain his professorial duties while tending to his burgeoning celebrity. "It's just becoming increasingly difficult to keep all the balls in the air at once."

Davies's 1996 book, *Are We Alone?: Philosophical Implications of the Discovery of Extraterrestrial Life*, argued that, in order to prove that life is not an accident but is intrinsic to the laws of the universe, it is worthwhile to search for extraterrestrial life. While he thinks it likely that life has evolved elsewhere in the universe, however, Davies does not believe that intergalactic travel is technologically viable—and is consequently skeptical of reported UFO sightings. He told Bill Broadway that it is "extremely unlikely that interplanetary travel takes place. It's dangerous and expensive. Why bother to do it if you can exchange information using radio?"

In *The Fifth Miracle: The Search for the Origin of Life* (2000), Davies analyzed modern scientific discoveries about the universe and theorized about the origins of humanity and the possible existence of other life in the universe. In 2002 Viking Press published Davies's *How to Build a Time Machine*, in which he asserted that time travel is indeed physically possible. "Traveling into the future is a technological problem, not a physics problem," Davies told Gabriella Boston for the *Washington Times* (February 28, 2002). "If we had the same dedication to building a time machine as we did to going to the moon in the 1960s, it might take 100 years." In 2010, Davies's book *The Eerie Silence: Renewing our Search for Alien Intelligence* was released.

Davies is a popular radio and television personality. He has hosted a series of science documentaries for BBC Radio 3—two of which became books, *The Ghost in the Atom* (1986) and *Desperately Seeking Superstrings: A Theory of Everything?* (1992), the latter of which earned him a Glaxo Science Writer's Fellowship. In 1995 Davies hosted a six-part series entitled the *Big Questions*; the follow-up, *More Big Questions*, was aired on Australia's multicultural, multilingual Special Broadcasting Service (SBS) Television in 1998. In the series, Davies and the Australian interviewer Philip Adams discussed age-old questions regarding the universe and humanity's relevance in the grand scheme of things.

In 2001, Davies was awarded the Kelvin medal by Great Britain's Institute of Physics. His other awards and honors include the 1993 Advance Australia Award; inclusion among Australia's ten most creative people by the Sydney Bulletin in 1996; and a fellowship from Great Britain's Royal Society of Literature in 1999. Davies has also had an asteroid (known also by its number, 6870) named in his honor.

Davies, who officially became a citizen of Australia in September 2001, has two children, Annabel and Charles, with his wife, Susan

Woodcock Davies, and two stepdaughters, Victoria and Caroline, from Susan's previous marriage. In 2006, Davies moved from Australia to Arizona State University to establish the BEYOND Center for Fundamental Concepts in Science. In 2008, he was appointed co-director of Arizona State University's Cosmology Initiative, one of the largest cosmology programs in the United States. In his free time Davies enjoys studying World War II history, politics, economics, and geography.

## Further Reading

Broadway, Bill. "When Theology Meets Cosmology; Physicist Wins Prize for Work on Science and Spirituality." *Washington Post* 11 March 1995: B7. Print.

Dembart, Lee. "Book Review: Universe That Knows What It's Doing." *Los Angeles Times* 18 Mar. 1988: 16. Print.

Mirsky, Steve. "Paul Davies: Physics Could Help Fight Cancer." *Scientific American.* Nature America, 13 Apr. 2011. Web. 9 Aug. 2012.

Radford, Tim. "God and the New Physics by Paul Davies-Book Review." *The Guardian.* Guardian New and Media Limited, 16 March 2012. Web. 9 Aug. 2012.

# Edwards, Helen

## American physicist

**Born**: May 27, 1936; Detroit, Michigan

The particle physicist Helen Edwards is perhaps best known for having led the effort to build the Tevatron, which was the world's most powerful particle accelerator at the time, prior to the construction of the Large Hadron Collider in Switzerland. Edwards has received numerous professional honors in her career, including a 1989 National Medal of Technology; a 1988 MacArthur Foundation "genius" award; the Department of Energy's 1986 Ernest O. Lawrence Award, "for her leadership in the construction and commissioning of the Tevatron at Fermilab"; and a 1985 Prize for Achievement in Accelerator Physics and Technology, awarded by the US Particle Accelerator School for her "essential contributions in making the world's first superconducting synchrotron a reality." Edwards served as a guest researcher at Fermilab, the US Department of Energy's particle physics laboratory near Batavia, Illinois, and at the Deutsches Elektronen Synchroton, in Germany. She is a fellow of the American Physical Society and a member of the National Academy of Engineering.

### Early Life and Education

The daughter of Edgar and Mary Thom, Helen Edwards was born Helen Thom in Detroit, Michigan, on May 27, 1936. She grew up in the city's suburbs, then attended an all-girls school near Washington, DC. By her own recollection, Edwards displayed an aptitude for the sciences—particularly engineering—at an early age. "In school, grade school, and junior high," she told the Virtual Museum of Virginia Tech in an online interview, "I was lousy at spelling and lousy at reading. It took me forever to read, but I found I enjoyed mechanical tasks, figuring out how things were put together and how they worked; I also enjoyed nature, biology, and the natural sciences."

Edwards attended Cornell University in Ithaca, New York, and received a bachelor's degree in physics in 1957. As one of two women among the fifteen physics majors in her graduating class, Edwards has said she occasionally encountered the attitude that "educating women in the sciences was a waste of time because they would choose

> **"It took me forever to read, but I found I enjoyed mechanical tasks, figuring out how things were put together and how they worked."**

marriage over a research career." But she feels that her physics professors—in particular Boyce McDaniel, who later served as faculty adviser for her doctoral thesis—were extremely supportive of her intellectual development. "Those guys were immensely helpful," Edwards recalled. (She has since endowed a chair in accelerator physics at Cornell in honor of McDaniel.) At Cornell, Edwards also met her future husband, the particle physicist Donald A. Edwards.

## Life's Work

For graduate school Edwards remained at Cornell, earning a master's degree in physics in 1963 and a PhD in the same subject in 1966. Thereafter, she served as research associate for four years at Cornell's Laboratory for Nuclear Studies, where she worked with the university's particle accelerator. In 1970 Robert Wilson, then the director of Fermilab and formerly the laboratory director at Cornell, invited Edwards to join him in Illinois as Associate Head of the Booster Group. Edwards retained that position through 1987, when she was named Head of Fermilab's Accelerator Division. While at Fermilab, Edwards made her contribution to building the Tevatron. In 1989 Edwards left Fermilab to help construct the Superconducting Supercollider, a particle accelerator that would have been housed in a fifty-four-mile tunnel outside Dallas, Texas. (After spending $2 billion on the collider, Congress scrapped the project in 1993.) In 1992 Edwards returned to Fermilab as a guest scientist.

Like all particle accelerators, the Tevatron is essentially an extremely powerful microscope—an instrument that allows scientists to probe the minuscule world of subatomic elementary particles. It works by first accelerating streams of protons and their anti-matter companions, anti-protons, to high velocities (close to the speed of light) inside of a large circular structure (the Tevatron is four miles in circumference), then causing the two groups of particles to collide head on. While the energy generated by this collision can produce a wide variety of sub-atomic particles—physicists sometimes speak of a "subatomic zoo"—most of these are extremely short-lived and quickly decay into other particles. Yet, by carefully examining the "debris"—as recorded by the trails that charged particles leave on plates of photographic film, for example—it is possible to deduce the existence and properties of extremely transient, extremely small particles. (Scientists have used Fermilab accelerators to experimentally detect two major components of what they call the "Standard Model of Fundamental Particles and Forces: the bottom quark, discovered in 1977, and the top quark, detected in 1994.) In September 2001, the Tevatron was closed due to budgetary constraints and the completion of the Large Hadron Collider, which has replaced the Tevatron as the most powerful particle accelerator in the world.

## Further Reading

"Fermilab's Helen Edwards Receives Prestigious 2003 Robert R. Wilson Prize from the American Physical Society." *Fermilab Press Pass.* U.S. Department of Energy, 22 Oct. 2002. Web. 9 Aug. 2012.

"Helen Edwards is Associate Head of Booster Section." *The Village Crier* 2.11 (19 Mar. 1970). Print.

"MacArthur Foundation Names 31 Recipients of 1988 Awards." *New York Times* 19 July 1988: A23. Print.

Pappas, Stephanie. "Farewell, Tevatron: Giant Atom Smasher Goes Silent After 28 Years." *LiveScience.* TechMediaNetwork.com, 30 Sept. 2011. Web. 9 Aug. 2012.

# Emanuel, Kerry A.

## American meteorologist

**Born**: April 21, 1955; Cincinnati, Ohio

During the last week of August 2005, a huge hurricane of near-record-breaking power slammed into the United States along the north-central coast of the Gulf of Mexico, bringing torrential rains and fierce winds to a 90,000-square-foot area encompassing parts of several southern states. Among the places hardest hit by the storm, named Katrina, was the city of New Orleans, Louisiana, where breaches in the levees along Lake Pontchartrain led to floods that left 80 percent of the city under water; more than a thousand people were killed, and many thousands were left homeless. An uncannily accurate description of what happened in New Orleans appears in a book that was published less than two months before Katrina began to form: *Divine Wind: The History and Science of Hurricanes*, by Kerry A. Emanuel, a professor of meteorology at the Massachusetts Institute of Technology. On August 4, 2005, less than four weeks before the deadly storm reached its peak strength, the British journal *Nature* published an article in which Emanuel reported that in recent decades both the intensities and the lifetimes of tropical cyclones (which, like "typhoons," is another name for hurricanes) had increased significantly. That phenomenon, Emanuel wrote, "is highly correlated with tropical sea surface temperature, reflecting well-documented climate signals"—one of them being global warming. "My results suggest that future warming may lead to an upward trend in tropical cyclone destructive potential, and—taking into account an increasing coastal population—a substantial increase in hurricane-related losses in the twenty-first century." Although, for years, many scientists in addition to Emanuel had warned of the dangers that unusually forceful hurricanes posed to New Orleans, Emanuel's prescience regarding the events in that city, along with his timely words concerning the link between global warming and hurricane intensity, made him a highly sought-after hurricane expert during and after Katrina's devastating rampage, and he was interviewed and quoted widely by the media. In 2006, *Time* magazine named him one of the "100 people who shape our world" (May 8, 2006). Earlier, in 1986, the American Meteorological Society honored him with its Meisinger Award for his work on atmospheric motions; in 1992 the organization presented him—along with Richard Rotunno, a senior scientist at the National Center for Atmospheric Research—with its Banner I. Miller Award, for their paper on the relationship between sea-surface temperature and the inner dynamics of hurricanes, which the society deemed to be an "outstanding contribution to the science of hurricane and tropical-weather forecasting." In addition to *Divine Wind*, Emanuel's professional writings include some 120 journal articles and book chapters and the book *Atmospheric Convection*, published by Oxford University Press in 1994. With three others, he coedited *Tropical Cyclone Disasters*, published by the Peking Press, also in 1994.

## Early Life and Education

The second of three brothers, Kerry Andrew Emanuel was born in Cincinnati, Ohio, on April 21, 1955, to Albert Emanuel II and Marny Catherine (Schonegevel) Emanuel. His mother was a former flight instructor and homemaker, and his father was an aircraft mechanic. During his childhood his family moved several times, setting up house in Pennsylvania, Maine, and Florida as well as Ohio. The few hurricanes the Emanuels experienced while they lived in Florida made an indelible impression on young Kerry. Early on, the subjects of climate and weather interested him more than any others. In school, Emanuel told *Current Biography*, he was "pretty much a straight arrow." He attended the Massachusetts Institute of Technology (MIT) in Cambridge, Massachusetts, where he earned an bachelor's degree in earth and planetary sciences in 1976 and a PhD in meteorology only two years later. From 1978 to 1981 Emanuel taught as a member of the Department of Atmospheric and Oceanic Sciences at the University of California, Los Angeles (UCLA). Concurrently, in 1979, he held a postdoctoral fellowship at the Cooperative Institute for Mesoscale Meteorological Studies at the University of Oklahoma. (Mesoscale meteorology is the study of thunderstorms and other weather systems that extend anywhere from about

10 miles to about 600 miles.) In 1981 he returned to MIT as an assistant professor in the Center for Meteorology and Physical Oceanography, which is associated with the Department of Earth, Atmospheric, and

> **"I was trying to convey a sense of hurricanes as not just things of scientific interest, but as beautiful. A leopard is a very beautiful animal. But if you took it out of its cage, it would go for your jugular. Anyone can understand that neither a leopard nor a hurricane is a willful killer."**

Planetary Sciences; he was promoted to associate professor in 1983 and to full professor in 1987. He directed the center from 1989 to 1997.

## Life's Work

Emanuel began his research on hurricanes in the early 1980s, after he was assigned to teach a course in tropical meteorology. While he was preparing his notes for the class, as he recalled to Claudia Dreifus for the *New York Times* (January 10, 2006), "I realized I didn't understand what I'd been taught on the subject. As with many things, you think you understand something until you try to teach it. After some reading, I realized that the reigning theory [of hurricane formation] had to be wrong." A hurricane forms when water vapor evaporates from the ocean and later condenses as a ring of tall thunderstorm clouds surrounding the eye of the storm. The accepted theory of how hurricanes develop, known as conditional instability of the second kind, as Emanuel told Dreifus, "held that the main thing that drives a hurricane is just ingestion of enormous quantities of water vapor from the atmospheric environment"; thus, it would depend on a "magical grouping" of winds and clouds. But the predictions that Emanuel derived from the theory were false, he discovered. "So it became a very big intellectual challenge to me. The more I got into it, the more interesting it became." Emanuel amended the theory of hurricane development in a way that accounted for the connection that exists between hurricane circulation and evaporation from the surface of the sea.

Hurricanes gather strength from the heat energy in tropical oceans. Warmer ocean waters evaporate more quickly than cooler ones and have been linked to hurricanes that are more powerful and travel faster than others. In a paper for *Nature* (April 2, 1987) entitled "The Dependence of Hurricane Intensity on Climate," Emanuel suggested that global warming, which causes the surface water of oceans to get warmer, could lead to an increase in hurricane strength. In preparing his paper, Emanuel assumed that if the quantity of carbon dioxide (one of the so-called greenhouse gases) in the atmosphere doubles by sometime in the twenty-first century, the temperature of the surface waters of the oceans in tropical regions will rise by about 2 to 3 degrees Celsius, or 4 to 5 degrees Fahrenheit, and, in turn, "the maximum possible intensity for hurricanes could rise 40 percent to 50 percent generally and 60 percent in the Gulf of Mexico," as Malcolm Ritter wrote for the Associated Press (April 1, 1987), paraphrasing Emanuel's findings. But the assumption regarding the relationship between carbon-dioxide increase and water-temperature increase could prove to be unfounded, as Emanuel told Ritter. Indeed, until recently Emanuel believed that nobody had found any evidence that hurricanes were growing more intense; as Claudia Dreifus wrote, he remained a "cautious centrist on questions of global warming and hurricane ferocity," often asserting that "no firm link had been established between warming and the intensity and frequency of hurricanes." In an interview with Jules Crittenden for the *Boston Herald* (November 29, 1998), for example, Emanuel suggested that, even if projections of global warming were accurate, no evidence pointing to the increasing intensity of hurricanes would be findable until the last quarter of the twenty-first century. "In the Atlantic, natural variability is what we need to worry about right now," he said, referring to an apparent twenty-year cycle of waxing and waning of hurricane strength, known as the Atlantic multidecadal oscillation.

Then, in the early 2000s, while conducting research on El Niño (an irregularly occurring current of unusually warm water in the Pacific Ocean) and other phenomena related to climate oscillations, Emanuel unintentionally discovered that hurricanes in the North Atlantic and western North Pacific had increased in overall power by roughly 60 percent since the 1970s. Until then, he told Dreifus, he and others had predicted that "if you warmed the tropical oceans by a degree Celsius, you should see something on the order of a 5 percent increase in the wind speed during hurricanes. We've seen a larger increase, more like 10 percent, for an ocean temperature increase of only one-half degree Centigrade"—a trend closely correlated with the increase in ocean and air temperatures due to global warming. Emanuel acknowledged the existence of that correlation in *Nature* (August 4, 2005), thus indicating how much his thinking about hurricane intensities had changed by that time.

Critics of Emanuel's *Nature* paper, which drew upon data from some 4,800 hurricanes, charged that he had left out measurements of storms in the 1950s and 1960s because they were inconsistent with his findings, and thus that he had based his conclusions on incomplete historical data. "His conclusions are contingent on a very large bias removal that is as large or larger than the global warming signal itself," Christopher Landsea, a researcher at the federal National Oceanic and Atmospheric Administration (NOAA) in Miami, Florida, told Joseph B. Verrengia for the Associated Press (August 1, 2005). Landsea, along with two others who disagreed with Emanuel—Colorado State University meteorologist William Gray and the director of the National Hurricane Center, Max Mayfield—blamed the rise in hurricane intensity since the 1970s on the Atlantic multidecadal oscillation. "That cycle," according to Gray, "caused eleven major hurricanes to hit Florida's peninsula from 1933 to 1965," as Robert P. King wrote for the *Palm Beach (Florida) Post* (October 8, 2005). "In contrast, just one major hurricane—Andrew in 1992—struck during the lull that followed for the next three decades." Mayfield insisted before a US Senate Committee that increased hurricane activity was due to the natural cycle and not to global warming. Emanuel responded to criticisms of his paper by reaffirming his stance. "I maintain that current levels of tropical storminess are unprecedented in the historical record and that a global-warming signal is now emerging in records of hurricane activity," he wrote in a letter to *Nature* (December 29, 2005), as quoted by Mark Schleifstein in the New Orleans *Times-Picayune* (July 9, 2006). Less than six months later, Emanuel told Marc Airhart for *Earth and Sky* (June 14, 2006), "I think the idea that [the increase in hurricane intensity is] part of a natural cycle is dead."

Much of the media coverage of Emanuel's work during and after the Katrina disaster oversimplified his findings. In particular, news stories often failed to mention that his data came from records of hurricanes that had remained over the ocean as well as the much smaller number that had made landfall. Moreover, reporters neglected to remind the public that no conclusion about the connection between global warming and hurricane intensity can be derived from measurements of a single storm or even from a fraction of all hurricanes—although, as Jeffrey Kluger pointed out in *Time* (May 8, 2006), it was not surprising that journalists reporting on the events in New Orleans did just that. "It's easy to argue about the hypothetical causes and effects of global warming," Kluger wrote. "It's a lot harder for any serious disagreement to continue when extreme weather is demolishing a major city." "If you consider hurricanes over their entire life and not just when they make landfall, you really do see an upward trend in the power of hurricanes, not in their frequency but in the magnitude of the wind speed and also in their duration," Emanuel told Alex Chadwick for the National Public Radio program *Day to Day* (September 6, 2005). "But you really can't see such tendencies in landfalling storms in the US simply because their numbers are so small. . . . It's impossible statistically to detect any kind of meaningful trend in that."

Emanuel's book *Divine Wind: The History and Science of Hurricanes*, which was published on July 1, 2005, was judged by most reviewers to be well-written, beautifully illustrated (with images of works of art as well as photos of storms), and suitable for lay readers as well as scientists. "Connoisseurs of natural disasters will devour [the book], but I hope it finds its way into numerous school libraries, too," Karen R. Long wrote for the *Cleveland, Ohio, Plain Dealer* (September 11, 2005). "The author's gifts as a science teacher combine with a cultivated taste in folklore, literature, and art." She also wrote, in mentioning Emanuel's description of the hypothetical sequence of events if a Category 5 hurricane were to hit New Orleans: "The only thing

[Emanuel] left out . . . is the likelihood of a loose barge crashing into the levee." In a review for *Weatherwise* (July/August 2006), Jeffrey B. Halverson wrote, "Those who have been touched by the awesome power of hurricanes, are curious about storms' long association with world culture, or just plain confused by what is being presented on the evening news will benefit from reading *Divine Wind.*" When Dreifus asked Emanuel how he felt about the timing of the book's publication, he said, "Not terribly good. If one is just interested in sales, I suppose it was fortuitous. But I was trying to convey a sense of hurricanes as not just things of scientific interest, but as beautiful. A leopard is a very beautiful animal. But if you took it out of its cage, it would go for your jugular. Anyone can understand that neither a leopard nor a hurricane is a willful killer."

## Climate Change and Hypercanes

A decade earlier, in a paper for the *Journal of Geophysical Research* (July 20, 1995) cowritten by Richard Rotunno and three others, Emanuel wrote about the types of hurricanes that would form if an asteroid were to hit a body of warm, tropical waters on Earth, or in the event of "shallow-sea volcanism, or, possibly, by overturning of superheated brine pools formed by underwater volcanic activity." The computer simulations that he and his colleagues devised indicated that the resultant groups of storms, dubbed "hypercanes," could produce winds of up to 90 percent the speed of sound. (The speed of sound varies with the type of medium through which the sound is moving—for example, air or water—and the temperature of the medium. Under "standard" conditions at sea level, the speed of sound is about 761 miles per hour.) Moreover, unlike normal hurricanes, hypercanes could drastically affect the climate of the entire planet. "[Emanuel's] model predicts that a hypercane would form in about a day if an area of water only 50 kilometers [about 31 miles] in diameter is heated to a temperature of 50 degrees Celsius [122 degrees Fahrenheit]," Jeff Hecht wrote for *New Scientist* (February 4, 1995). "He calculates that such a hot spot could be formed if an object larger than 10 kilometers in diameter [about 6 miles] crashed into a shallow sea." In addition to intense winds, the hurricane could have an eye with atmospheric pressure less than a third the normal level. Air at such low pressure would act like a vacuum, pulling dust and water vapor up into the stratosphere, where it could remain for years. Moreover, as Hecht reported, "The sun's ultraviolet radiation would act on stratospheric water droplets to form hydroxyl radicals. Together with chlorine from saltwater thrown up by the hypercane these radicals could destroy vast quantities of stratospheric ozone. UV radiation would then reach the Earth's surface unhindered, killing organisms on land and in the upper layers of the ocean." Emanuel believes that hypercanes could have contributed to the extinction of the dinosaurs.

In his 2007 book, *What We Know About Climate Change*, Emanuel continued to warn of the potential dangers of global warming as it relates to intensified hurricane activity. He offered a scientific explanation of global warming and discussed the media's influence on the public's understanding of it. In an assessment for NYBooks.com (October 11, 2007), Bill McKibben wrote, "In an epic feat of concision, [Emanuel] manages in eighty-five very small pages to explain the state of the science of climate change, concluding on the optimistic note that 'the extremists [who deprecate the threat of climate change] are being exposed and relegated to the sidelines, and when the media stop amplifying their views, their political counterparts will have nothing left to stand on.'"

When Marc Airhart asked Emanuel if his new role of public hurricane expert had changed his life, the scientist answered, "It has. A lot of people write to me or phone me. They're all completely well-meaning people and they're curious about nature. It's frustrating, because I'd dearly like to be able to talk to each of these people and answer their emails, and it's become humanly impossible for me to do that. So it's forced me to make judgment calls. On the one hand, I want to use the opportunity to convey to the public my understanding of the problem. That's very much a part of the duty, perhaps even a moral obligation, of a scientist. But on the other hand, I want to get back to the work, to the research and the teaching. It's hard to know where to strike that balance under circumstances like this."

Emanuel lives in Lexington, Massachusetts, with his wife, Susan Boyd-Bowman, whom he married in 1990, and their son, David Tristan Emanuel.

## Further Reading

Dreifus, Claudia. "With Findings on Storms, Centrist Recasts Warming Debate." *New York Times* 10 Jan. 2006: F2. Print.

Kluger, Jeffrey. "The 2006 TIME 100: Kerry Emanuel." *TIME Magazine* 8 May 2006: 92. Print.

Rittner, Don. *A to Z of Scientists in Weather and Climate*. New York: Facts on File, 2003.

# Ericsson-Jackson, Aprille J.

## American aerospace engineer

**Born**: April 1, 1963; Brooklyn, New York

Aprille J. Ericsson-Jackson, an aerospace engineer at NASA's Goddard Space Flight Center (GSFC) in Greenbelt, Maryland, has the distinction of being not only the first African American woman to receive a PhD in mechanical engineering from Howard University in Washington, DC, but also the first American to receive a doctorate in that field with an aerospace specialization. At GSFC she works to ensure that spacecraft will perform well during NASA missions. Ericsson-Jackson has also worked hard to increase diversity in the sciences, regularly visiting schools as a member of both Women of NASA and GSFC's Speakers Bureau and encouraging young women and minority students to pursue careers in scientific fields.

### Early Life and Education

Aprille J. Ericsson-Jackson was born on April 1, 1963, in the New York City borough of Brooklyn. Her parents separated when she was eight, and her mother raised Aprille and her two sisters in Brooklyn's Roosevelt housing projects. Ericsson-Jackson discovered her talent for math and science while attending Marine Park Junior High School, where she was the only black student in the Special Progress program, which included instruction in mathematics, earth science, biology, and chemistry. Thanks in part to the encouragement and support of her mother, she achieved outstanding scores on state and city exams and once won second place in her junior high's science fair. She passed the entrance exams for all three of New York City's highly competitive public technical high schools. At the age of fifteen, she moved to Cambridge, Massachusetts, where she lived with her grandparents while attending the Cambridge School of Weston, a prestigious private college-prep school that gave her a full scholarship three years in succession. She enjoyed sports and proved to have a talent for football, basketball, and softball, playing on student teams.

In the summer after her junior year of high school, Ericsson-Jackson participated in UNITE (now known as Minority Introduction to Engineering, or MITE), a two-week program for African American students that inspired her to consider a career in aerospace engineering. "Over the course of the program, we were exposed to several engineering disciplines," she told Michael Baine for Space.com (September 22, 2000). "One was civil engineering, where we made small

---

**"And one of the things that's really sad is that teachers and educators lack Internet access, and their schools are missing out. Hopefully all that will change."**

---

bridges that were tested by loading until failure. That was my first attempt at designing anything as a pseudo-engineer. The director of the program was a biomedical engineer, which also interested me, but I decided I did not want to be a medical doctor because of the memorization needed and I did not want to do civil engineering/architecture because buildings and structures did not move." Ericsson-Jackson also visited an air-force base in New Hampshire, where her performance at the controls of a flight simulator indicated a level of ability comparable to that of a pilot. She maintained excellent grades and scored high on her PSATs; she also found the time to volunteer as a physical-education instructor at a number of local elementary schools.

Ericsson-Jackson was accepted as a student at the Massachusetts Institute of Technology (MIT) in Cambridge. During her freshman year she attended an aerospace-engineering seminar in which she learned about "the different disciplines within aerospace," as she told Michael Baine. "It also allowed me to meet various faculty members in the department and exposed me to their research. . . . I kept up my grades and was accepted into the aero/astro program." As an undergraduate at MIT, Ericsson-Jackson worked on several projects related to manned space missions. In 1986 she received a bachelor's degree in aeronautical and astronautical engineering from MIT. With the help of a number of grants, she then attended Howard University as a graduate student in the Large Space Structures Institute. She also delivered technical papers in Germany, Canada, England, and throughout the United States. She earned both a master's degree (1990) and a PhD (1995) in mechanical engineering, making her the first African American woman to receive a doctorate in this discipline from Howard University.

## Life's Work

Ericsson-Jackson was hired by the National Aeronautics and Space Administration (NASA) in 1992. She works in the Guidance, Navigation and Control Center of NASA's Goddard Space Flight Center, where she concentrates on satellite projects such as the X-Ray Timing Explorer (XTE) and the Tropical Rain Forest Measurement Mission, testing spacecraft designs by conducting simulations of performance. She has worked on the Microwave Anisotropy Probe (MAP), a satellite designed to seek out clues to the origins of our galaxy, the Milky Way, by measuring the properties of cosmic microwave background radiation found in the sky. MAP launched on June 30, 2001, and, as Charles Bennett, the MAP Science Team's principal investigator, told *Current Biography*, it "is now successfully in its nominal orbit at the second Earth-Sun Lagrange point . . . a million miles from Earth." Bennett added, "All systems are working very well."

In an autobiographical statement Ericsson-Jackson sent to *Current Biography*, she wrote that in the future she hopes to work as "a mission specialist for the astronaut program, [an] Aerospace Engineering professor; and an advisor to the White House through the Office of Science and Technology Policy."

Believing that the Internet is a key component of disseminating information on technical careers to those who need it, Ericsson-Jackson has created an email list through which she provides information to those interested in learning more about educational and employment opportunities in the sciences. "The Internet can also bring new resources to the African American community," she told ZDNet (online). "In Washington, lots of African American kids use public computers to look for funding for college and to apply online. Who's to say that the corner bookie couldn't have earned an MIT degree if he had been given the opportunity? It costs money to be online, unfortunately, and a lot of people still can't afford the cost of a system that's fast enough, with a printer and everything else you need. And one of the things that's really sad is that teachers and educators lack Internet access, and their schools are missing out. Hopefully all that will change."

Ericsson-Jackson is a recruiter for GSFC; she has applied to NASA's astronaut program, but has thus far been unable to participate due to an asthma condition and surgical repair of both knees. She has taught at Howard University's Department of Mechanical Engineering and is an adjunct professor in the Department of Mechanical Engineering and Mathematics at Bowie State University in Maryland, where she was hired to improve the engineering curriculum. She created two new courses for freshmen and sophomores at Bowie. Should she fulfill her goal of becoming a professor at Howard University, Ericsson-Jackson hopes to create and chair an aerospace department. She is a member of a number of organizations, including the American Astronautical Society, the American Institute of Aeronautics and Astronautics, the American Society of Mechanical Engineers, the Society of Women Engineers, and Sigma Xi. She also participates in many community-outreach programs and has visited the White House for meetings on issues related to science, engineering, and technology. In 1997 Ericsson-Jackson received the Women in Science and Engineering (WISE) Award, which recognizes the best female engineer in the federal government. The following year NASA's African American Awards Committee named her one of GSFC's outstanding African Americans. She received special recognition at the Black Engineer Awards ceremony sponsored by *U.S. Black Engineer and Information Technology Magazine*. iVillage.com included her on its list of "Women Who'll Rule" in the near future.

Ericsson-Jackson's many hobbies include reading, sports, woodworking, sewing, and baking. She lives in Washington, DC.

## Further Reading

"Dr. Aprille J. Ericsson." *Visiting Researcher Profile*. National Center for Earth and Space Science Education, Sept. 2007. Web. 9 Aug. 2012.

"Female Frontiers." *NASA Quest*. NASA. Web. 5 Apr. 2012.

# Gray, William M.

## American meteorologist

**Born**: October 9, 1929; Detroit, Michigan

"The best way to predict the future is to study the past," the meteorologist William M. Gray told Diane Levick for the *Hartford (Connecticut) Courant* (September 17, 1999). "There are surprising precursor signals that tip off what the coming season will be like." Gray was referring to the hurricane season; long before he spoke to Levick, he had become one of the world's foremost authorities on hurricanes and other tropical storms: how they form, their structure, the paths they take, their frequency, and global factors affecting their intensity. Gray's intensive study of meteorology began during his service in the Korean War, in the early 1950s. Later, in graduate school at the University of Chicago, he became a protégé of Herbert Riehl, "widely regarded as the father of tropical meteorology," according to *Meteorology and Atmospheric Physics* (March 1998). Riehl became Gray's research partner as well as mentor, and after Riehl moved to Colorado State University, Gray followed him in 1961. Gray has worked there ever since; hundreds of professional meteorologists who hold jobs at the National Hurricane Center and elsewhere were students of his. For the last few years, he has held the title of professor emeritus of atmospheric sciences at the university.

Although he has repeatedly reminded the public that there is still no foolproof way to anticipate weather conditions with precision, in 1984 Gray began to make annual predictions regarding each year's hurricane season, and he has never been wide of the mark. His predictions have appeared in hundreds of newspapers and on many websites. Reasonable accuracy in forecasting is invaluable to farmers, the most obvious example of people who depend directly on the weather for their livelihoods. Others who depend on accurate weather forecasts are government agencies whose mission is to help people whose lives or property may be imminently endangered by severe weather;

companies that insure against weather-related damage to property; and people who engage in futures trading of grains and other agricultural commodities. Millions, if not billions, of ordinary people in the developed world have come to listen to or look at weather reports as a daily ritual, and they, too, rely on the information provided.

In recent years meteorologists—those who study Earth's atmosphere and monitor and forecast the weather—have expanded their field of expertise to the investigation of global warming and its effects. Gray, too, has studied global warming, but unlike most of the scientists whose views have been made public on that subject, he believes that human activities—foremost among them the burning of fossil fuels—have had and will have negligible effect on the increases in the temperatures of the air and the surfaces of the oceans; rather, he believes that such changes constitute evidence of the next in the series of Earth's natural cycles of atmospheric warming and cooling. He has frequently labeled the linking of human activities and global warming as a "hoax," as quoted by Nicholas Riccardi in the *Los Angeles Times* (May 30, 2006). He said to Alan Prendergast for WestWord.com (June 29, 2006), "I'm going to give the rest of my life to working on this stuff."

## Early Life and Education

William Mason Gray was born on October 9, 1929, in Detroit, Michigan, and grew up in Washington, DC. As a young man he loved baseball and was a pitcher of some promise; an injured knee prevented him from pursuing the sport professionally. Gray attended George Washington University; he earned a bachelor's degree in geography in 1952. The Korean War was about to enter its third year then; wary of being drafted into combat after college, Gray had preemptively signed up for a meteorological program in the Air Force that would pay for his graduate studies in climatology following three years of military service. During the first year of his service, he studied meteorology at the University of Chicago in Illinois. After his discharge from the Air Force, Gray enrolled in the graduate meteorology program at that school. There, he met Herbert Riehl. Gray began to help Riehl with his research, in the course of which they would hire pilots to fly them into or above tropical storms. In 1958, in one of the most memorable of such flights, they traveled in a converted bomber along the coast of North and South Carolina while the 150-mile-per-hour winds of Hurricane Helene—a Category 4 hurricane and the strongest of that year's tropical storms—raged below them. (Sustained wind speed is the measurement that determines a hurricane's category, with Category 1 being the least intense and Category 5 the most intense. Only two officially recorded Category 5 hurricanes have struck the US mainland: Camille in 1969 and Andrew in 1992.

Historical accounts show that humans have attempted to predict the weather for at least 2,500 years. People probably strived to make accurate long- and short-term forecasts for thousands of years before that, beginning when they turned from hunting and gathering to growing their own crops. The science of forecasting received a huge boost with the invention of the telegraph in the mid-1800s, which made possible the rapid dissemination of data across long distances. With the dawn of the computer age, meteorology again advanced exponentially, as computers enabled scientists to cumulate, analyze, and compare vast amounts of data rapidly. The third invaluable aid to the field was the invention of the weather satellite; the US government's first successful launch of such a satellite came in 1960. When Gray began working with Riehl, meteorology was still relatively primitive. "Those were the days of pre-satellite data, when the airplane was king and flew into storms," Gray told Deborah Mendez for the Associated Press (September 26, 1994).

Gray earned an master's degree in meteorology from the University of Chicago in 1959; his thesis was entitled "On the Balance of Forces in Hurricane Daisy of 1958." He completed a PhD in geophysical sciences in 1964, writing a dissertation entitled "On the Scales of Motion and Internal Stress Characteristics of the Hurricane." Earlier, from 1957 to 1961, he had worked as a research assistant at the University of Chicago, then followed Riehl to Colorado State University in Fort Collins to take the post of assistant meteorologist in the newly founded Department of Atmospheric Science. Gray became a full professor of atmospheric sciences in 1974.

## Life's Work

The research of Gray and Riehl focused on storms with paths in the Atlantic Ocean. For some time they made little progress in making sense of their data and thus becoming adept at forecasting such storms. The accuracy of their predictions improved when they began collecting data from further afield. "The problem was that we'd been

> **"The best way to predict the future is to study the past."**

looking locally," Gray told Alan Prendergast. "You had to look globally." He and Riehl identified several factors that influenced the patterns and strengths of hurricanes but occurred far from the East Coast of the United States. They included El Niño, the irregularly occurring warming of seawater in the eastern Pacific; sea-level air pressure in the Caribbean Basin; rainfall in West Africa; and the direction of stratospheric winds above the equator and above the lower Caribbean Basin. (Earth's atmosphere has five layers. The stratosphere is the second; it extends from about 31 to about 34 miles above Earth's surface.) Those factors, Gray told William Fox for the *St. Petersburg Times* (June 2, 1987), account for only about half the variability of hurricane seasons from one year to the next.

In meteorological jargon, a tropical storm has winds of between 34 and 74 miles per hour. Storms whose wind speeds are less than 34 miles per hour are called tropical depressions; those with wind speeds greater than 74 miles per hour are called hurricanes. (In other parts of the world, hurricanes are known as cyclones or typhoons.) In 1984 Gray publicly forecast, for the first time, how many hurricanes (seven) and tropical storms (ten) he expected to occur that summer. Five hurricanes developed, along with twelve storms. That prediction was sufficiently on target to attract some attention; more came his way after he improved his methods and launched the University of Colorado's Tropical Meteorology Project. In 1985 he predicted eight hurricanes, and seven occurred; in 1986, he predicted four, and four occurred;

in 1987 he predicted four hurricanes and three tropical storms, and there were three hurricanes and four tropical storms. Additionally, in 1987 he forecast that on thirty-five days a hurricane or tropical storm would be active during the Atlantic hurricane season (which, officially, extends from June 1 through November 30); the actual number was thirty-six. By the early 1990s Gray had made a name for himself as a trustworthy forecaster of hurricanes and tropical storms. The ease with which he won grants and other funding for his research reflected his success.

In his interview with William Fox in 1987, Gray noted that, because of global wind patterns, there had been a smaller-than-average number of intense storms since 1970. But he expected those patterns to change within the next dozen years or so, and then, he warned, people would need to "watch out." A decade later Gray testified before members of the Subcommittee on Housing and Community Opportunity of the US House of Representatives' Banking and Financial Services Committee regarding the possibilities of hurricanes making landfall in the United States. "Global observations indicate that we are entering a new era of increased intense or 'major' hurricanes," he said, as quoted by the Federal News Service (June 24, 1997). "If our interpretation of climate trends which are indicating increased intense hurricane activity bear out, then the cost of US hurricane-spawned destruction will be of a magnitude as never before seen." During his testimony he also denounced the theory of manmade global warming, declaring, "These are natural changes."

## Debate over Global Warming

In an interview with William K. Stevens for the *New York Times* (February 29, 2000), Gray said, "I don't think we're arguing over whether there's any global warming. The question is, 'What is the cause of it?'" According to Gray, rising temperatures are part of a natural cycle of heating and cooling caused by, among other factors, thermohaline circulation (THC), also called the global ocean conveyor. All the world's oceans are connected, and scientists have detected large-scale patterns in the movement of seawater at the surface and at all depths. "Overall, the globe has warmed over the past 120 years," Gray told

Alan Prendergast. "That's due, in my view, to a multi-century slowing of the thermohaline. We're coming out of a little ice age, and overall, the thermohaline has been slowing. But that doesn't mean it doesn't speed up for thirty or forty years."

In his conversation with Stevens, Gray also disputed the theory that global warming is connected with an increase in hurricanes. He later presented his evidence for that assertion, and for the role THC plays in climate change, in the paper "Global Warming and Hurricanes" that he presented in 2006 at the twenty-seventh Conference on Hurricanes and Tropical Meteorology, sponsored by the American Meteorological Society. He began his paper with a quote from the Republican US senator James Inhofe of Oklahoma, who declared on the floor of the Senate in 2003, "Global warming caused by human activity might be the greatest hoax ever perpetrated on the American people." Gray ended the paper by referring to the novel *State of Fear* (2004) by Michael Crichton. (One of the chief villains in *State of Fear*, the head of an environmental organization, conspires with ecoterrorists to convince the public of the danger of global warming so as to scare them into donating generously to his group; the hero debunks the global-warming theory, his arguments supported by footnotes in which Crichton cited papers by real-life global-warming deniers such as Gray.) As Gray noted, the American Association of Petroleum Geologists (AAPG) honored Crichton with its Journalism Award for *State of Fear*, a book that, he added, "dismisses global warming as a conspiracy." Gray quoted Larry Nation, an AAPG spokesperson, as saying about *State of Fear*, "It is fiction. But it has the absolute ring of truth."

Several of the climatologists who manage the website RealClimate. org (April 26, 2006) refuted "a few key points" from Gray's paper that "illustrate the fundamental misconceptions on the physics of climate that underlie most of Gray's pronouncements on climate change and its causes." ("Our discussion is not a point-by-point rebuttal of Gray's claims; there is far more wrong with the paper than we have the patience to detail," they explained.) "None of the assertions are based on rigorous statistical associations, oceanographic observations, or physically based simulations," they wrote. "It is all seat-of-the-pants stuff of a sort that was common in the early days of climate studies." The renowned physicist, oceanographer, and climatologist Stefan Rahmstorf, a professor at Potsdam University in Germany who specializes in the role of ocean currents in climate change, wrote the next day in a message to RealClimate.org (April 27, 2006) that he had met Gray at a conference in 1998 and had "a nice discussion" with him. Gray "described to me in very clear terms how the thermohaline ocean circulation had decreased and increased at various times during the twentieth century," Rahmstorf recalled. "I felt very embarrassed—here I was, a young scientist who had been working already for seven years on the THC, and I had never even heard of these changes—I had thought it was basically unknown how the THC had varied over the twentieth century! Obviously I had a major gap in my grasp of the scientific literature. When I returned home I immediately did an extensive literature search—and to my surprise I found no studies at all that supported the very assured claims made by Gray."

Rahmstorf and others who criticize his views, Gray complained to Prendergast, "want to put me on the fringe. My brain is fossilized. I'm an old curmudgeon who doesn't change with the times. They use anything they can against you." Gray himself has publicly accused scientists of knowingly carrying out flawed experiments, even lying, for the sake of their funding. "It bothers me that my fellow scientists are not speaking out against something they know is wrong," he told Steve Lyttle for the *Sydney (Australia) Morning Herald* (October 14, 2007). "But they also know that they'd never get any grants if they spoke out. I don't care about grants." During the administrations of President Bill Clinton, though, Gray continued to apply for federal grants. "I must have been turned down thirteen times," he told Prendergast, adding that other scientists "were getting money to run these big [computer] models, and they were bending their objectivity to get the money." Greg Holland, the head of the Mesoscale and Microscale Meteorology Division of the National Center for Atmospheric Research in Boulder, Colorado, who completed his doctoral work under Gray, told Prendergast, "Getting support for research goes through a well-established peer review. . . . The lack of funding for [Gray's] research is related to the quality of his research." Peter Webster, a

professor of meteorology at the Georgia Institute of Technology and
a former collaborator of Gray's, told Prendergast that he had recently
refused to participate in a public panel with Gray because he could not
extract from Gray a promise to stick to the science and refrain from
ad hominem attacks. "I certainly wasn't going to stand in public and
let Bill berate me," Webster said. Holland told Prendergast, "[Gray]
has always been a natural thinker, the sort of person who asks difficult
questions. The unfortunate difference now is the way personalities are
being brought into it—and the denigration of perfectly good scientific
techniques."

When Prendergast asked him to consider the possibility that human
activities will indeed cause irreversible damage to the environment,
Gray replied, "There's nothing we can do about it anyway. We're not
going to stop burning fossil fuels. China and India aren't going to stop.
The little amount you might be able to cut out is negligble, and it's
going to hurt the middle class." With the aim of devoting himself to
recording his ideas about global warming in a published article, Gray
passed most of his Tropical Meteorology Project responsibilities to
Philip J. Klotzbach, who earned a PhD under his guidance. (Before
he ended his teaching career, Gray served as an adviser to more than
seventy master's-degree and doctoral-degree students at Colorado
State University.) "When I am pushing up daisies, I am very sure that
we will find that humans have warmed the globe slightly, but that it's
nothing like what they're saying," he told Prendergast. "I just don't
want to die and leave all these loose ends."

Gray is a research fellow at the Independent Institute, whose mission, according to its website, is "to transcend the all-too-common
politicization and superficiality of public policy research and debate,
redefine the debate over public issues, and foster new and effective directions for government reform." His honors include the Jule G. Char-
ney Award from the American Meteorological Association (1994); the
Banner I. Miller Award from the Atlantic Oceanographic and Meteorological Laboratory, a division of the National Oceanic and Atmospheric Administration (1994); and the Neil Frank Award from the
National Hurricane Conference (1995). Gray is a widower. He has
adult children and lives in Fort Collins.

## Further Reading

"Gray and Muddy Thinking About Global Warming." realclimate.org. Web. 5 Apr. 2012.

Prendergast, Alan. "The Skeptic." *Denver Westword.* Westword.com, 29 June 2006. Web. 5 Apr. 2012.

Riccardi, Nicholas. "Eminence Grise of Hurricane Forecasting." *Los Angeles Times* 30 May 2006: A6. Print.

Stevens, William K. "Global Warming: The Contrarian View." *New York Times* 29 Feb. 2000: F1. Print.

Svitil, Kathy A. "Discover Dialogue: Meteorologist William Gray." *Discover Magazine* Sept. 2005: 15. Print.

# Greene, Brian

## American physicist

**Born**: February 9, 1963; New York City, New York

Since Albert Einstein proposed the idea, physicists have been trying to come up with a "theory of everything." In Einstein's view, such a theory would be "a single master framework that would describe physics out to the farthest reaches of the cosmos and down to the smallest speck of matter," as the physicist Brian Greene put it in *Natural History* (February 2000). Greene and other contemporary physicists think about such a theory in terms of the four fundamental forces that exist in the universe: gravity, electromagnetism, and the two forces that are at work within atoms—the weak nuclear force, which causes radioactivity, and the strong nuclear force, which binds protons and neutrons. Thus, according to modern-day physicists, the so-called unified theory would, in Greene's words, "show all four forces to be distinct manifestations of a single underlying force" and, in addition, would "establish a rationale for the presence of the particular species of apparently fundamental particles"—quarks and leptons and their antiparticles, which make up the more than one hundred subatomic particles that have been identified so far. The "theory of everything," many physicists have proposed, will eventually be constructed by means of string theory. At its most basic level, string theory postulates that both atoms and the four fundamental forces are made up of small looplike entities—dubbed strings—which vibrate at different frequencies, and that the differences in frequencies distinguish subatomic particles from one another.

Brian Greene is one of the world's leading string theorists. He has also helped explain the complex world of string theory to laypeople by means of genial lectures and a bestselling book, *The Elegant Universe* (1999), which was a finalist for the Pulitzer Prize in nonfiction. "The ideas that we're currently investigating in cutting-edge physics have deep implications for questions that people have struggled with for thousands of years," Greene told an interviewer for *Columbia* (Fall 1999). "Questions that are really at the heart of the human condition, such as, What is time? What is space? What is the universe made of? It may well be that we are taking the final step to getting the deepest scientific answers to some of these questions. And it is imperative, I think, that everyone be able to share in this journey."

## Early Life and Education

The son of Alan and Rita Greene, Brian Greene was born on February 9, 1963, in New York City. His father, a one-time vaudeville performer, later worked as a voice coach and composer. The family lived on the Upper West Side of Manhattan, near the American Museum of Natural History and the Hayden Planetarium (now part of the museum's Rose Center for Earth and Space). Unlike many children, as Greene recalled to Alden M. Hayashi for *Scientific American* (April 2000), he "never really [felt] excited" by the museum's dinosaur exhibits, but he was fascinated by what he saw in the planetarium. "Ever since I can remember," Greene told Hayashi, "I was always questioning what the universe was made of and how it got to be the way it got to be." At a very young age, Brian showed a talent for arithmetic. When he was five, he would multiply thirty-digit numbers that his father would write out for him, doing the calculations on big sheets of construction paper that he and his father had joined with tape. "It's a weird thing, mathematics," Greene told Shira J. Boss for *Columbia College Today* (September 1999). "Even as a six-year-old you can learn a few rules and then play around with it. You can't do that with literature, where you need years of experience to say anything interesting."

When Greene was in sixth grade at Intermediate School 44 in Manhattan, his math teacher, knowing that he had exhausted the school's math resources, suggested that he go to Columbia University to find a tutor. Going from door to door in the computer-science department with his older sister, he had no luck until he ran into Neil Bellinson, a graduate student in mathematics, who agreed to tutor him for free. Greene was still getting instruction from Bellinson when he entered Stuyvesant High School, an elite New York City public school that accepts students on the basis of an exam. While at Stuyvesant he wrestled on the school varsity team, won citywide math competitions four years in a row, and was a finalist in the prestigious, nationwide Westinghouse Science Talent Search contest. Meanwhile, he had become deeply interested in mathematics as a tool for learning more about the universe. In an interview for SuperStringTheory.com, Greene said, "I think as an adolescent I had many of the questions and

> **"The ideas that we're currently investigating in cutting-edge physics have deep implications for questions that people have struggled with for thousands of years."**

concerns that many adolescents do . . . what's it all about, why are we here, what are we meant to be doing with our time and so forth. And it just occurred to me that many people much smarter than I had thought of these questions through the ages and come up with various solutions, none of which I guess were completely satisfying, and it didn't seem to me that I was going to come up with a solution to those particular problems. But it seemed to me that if one could gain a deep familiarity with the questions, a real profound understanding of the questions themselves—that is, why is there space, why is there time, why is there a Universe—then at least that would be the first step towards coming to answers."

In 1980 Greene entered Harvard University in Cambridge, Massachusetts, where he majored in physics. At Harvard he became aware of one of the great mysteries of modern physics, which is that the two major theories that describe the physical universe—the theory of relativity, which explains astronomical phenomena, and quantum mechanics, which describes the atomic and subatomic worlds—are mutually incompatible. "Quantum mechanics is based on the idea that things can flutter, things fluctuate," Greene Greene told Kai Wu, Justin Vazquez-Poritz, and Mike Wisz for the Cornell University campus publication *Science and Technology* (Winter 1996). "Even if they are in the lowest possible state of energy, even if they're resting on the table, they're still vibrating. And, if you try to combine that with general relativity, you run into big problems, because general relativity wants to model the universe as this very smooth, placid, varying space-time geometrical shape, and that's at odds with the frenetic flutter and jitter of quantum mechanics."

After earning a bachelor's degree, Greene enrolled at Oxford University in England as a Rhodes scholar. In 1985 he attended a lecture at Oxford about the "theory of everything." The focus was string theory, which purportedly reconciles the theory of relativity with quantum mechanics. Soon, Greene and fellow graduate students were learning as much as they could about the theory, sharing what they found out on their own. According to string theory, the universe is composed of tiny strings, or energy loops, which vibrate at different frequencies, thus creating electrons, photons, and other particles. "String theory comes along and says, do not think about elementary particles as being little points," Greene told the *Science and Technology* interviewers. "Think about them as being little loops. And when you do that, it turns out that you get exact meshing of gravity and quantum mechanics." However, string theory postulates that the universe has more than three spatial dimensions. Specifically, string theorists maintain that there are seven additional dimensions of space. These extra dimensions, according to one approach, are too small to be observed, as they are "curled up" very tightly. For that reason, string theory remains purely hypothetical. "We've been struggling for many years now to understand how it's possible that this theory that predicts ten dimensions can still be describing our world," Greene told the *Science and Technology* interviewers. Completely engrossed by string theory, Greene chose as the subject of his doctoral thesis, as Hayashi explained in *Scientific American*, "a possible way to coax experimentally testable predictions from string theory." He received a PhD from Oxford in 1987.

## Life's Work

Greene returned to Harvard for postdoctorate work before accepting a job teaching at Cornell University in Ithaca, New York, where his research focused on quantum geometry, the study of the mathematical properties and physical implications of the extra seven spatial dimensions. In the early 1990s, working with colleagues, Greene made two important discoveries. With physicist Ronen Plesser, he discovered, as Jennifer Senior reported in *New York* (February 1, 1999), "that for every possible shape of the cosmos there is a 'mirror shape' that generates an alternate universe with exactly the same properties." Then, in 1992, while on a sabbatical leave from the university, Greene demonstrated, along with Paul S. Aspinwall and David R. Morrison of Duke University, that if string theory is correct, the spacial "fabric" can tear and repair itself, and thus the universe can reshape itself. This conclusion, though purely theoretical, was important nevertheless, because the theory of general relativity prohibits ruptures in space-time. Greene and his fellow string theorists were eagerly awaiting the construction of the Large Hadron Collider in Geneva, Switzerland, which smashes protons with unprecedented power. The creation, by that method, of particles whose existence is predicted by string theory (such as the Higgs boson) would go far toward proving the validity of the theory. "What has drawn me to science is the thrill of discovery," Greene told Hayashi. "There's nothing like that moment of realizing that you've discovered something that has not yet been previously known."

In 1996 Greene left Cornell for Columbia University in New York City, where he started a string-theory program while teaching as a professor of physics and mathematics. "He has this great reputation," Greg Langmead, a graduate student, told Jennifer Senior. "Even among students who haven't taken anything with him. He's a great communicator, he's charismatic. He's clearly top-of-the-heap intellectually. So the fact that he has gobs of raw physical appeal on top of that—it gives him a really serious mystique." In February 1999, after two years of work, Greene's book, *The Elegant Universe: Superstrings, Hidden Dimensions, and the Quest for the Ultimate Theory*, was published; it spent more than four months on the *New York Times* bestseller list, while briefly becoming Amazon.com's highest-selling book. In a review of *The Elegant Universe* for the *Washington Post Book World* (March 7, 1999), the science journalist Marcia Bartusiak wrote, "Greene does an admirable job of translating a wholly mathematical endeavor into visual terms. Throughout his work, he writes with poetic eloquence and style." She also warned that Greene's "desire to reach the general reader may be overly ambitious. His discussions of gauge symmetries and

Calabi-Yau geometries will be best appreciated by the science-minded who seek an insider's perspective on the cutting edge of physics." In addition to being a finalist for the Pulitzer Prize, in June 2000 *The Elegant Universe* won the Aventis Prize, England's top honor for a science book. Greene went on a nationwide book tour to publicize *The Elegant Universe*. During the tour he lectured at the Guggenheim Museum in New York City to the accompaniment of the Emerson String Quartet, in an event that was called "Strings and Strings." The lecture/performance proved so popular that it was repeated, and another performance is being planned at Lincoln Center. In the fall of 1999, Greene was featured in a segment of *Nightline in Primetime: Brave New World*, on ABC-TV. Also in 1999, Greene and his colleagues received a $2.5 million grant from the National Science Foundation to restructure high-level mathematics and physics courses at Columbia. Greene also worked on a three-part series on unified theories for PBS's *Nova*, which was nominated for three Emmy Awards and won a Peabody Award. In 2005 Greene's second book, *The Fabric of the Cosmos: Space, Time and the Texture of Reality*, spent nearly six months on the *New York Times* bestseller list. With his wife, Tracy Day, Greene cofounded the World Science Festival in 2008, an "annual celebration and exploration of science," according to the event website. The *New York Times* has hailed the festival as a "new cultural institution." Greene also published a children's book about Albert Einstein and the theory of general relativity, called *Icarus at the Edge of Time* (2008). His most recent work, *The Hidden Reality: Parallel Universes and the Deep Laws of the Cosmos*, was released in 2011.

Greene has continued to investigate questions of time and space from the perspective of quantum mechanics. According to string theory, time and space may be manifestations of more complex entities. He and other physicists hope to discover why movement in space can be forward or backward, but movement in time appears to be possible only in one direction. Working in a different sphere, Greene helped the actor John Lithgow with scientific dialogue for Lithgow's role on the NBC-TV series *Third Rock from the Sun*, and he had a cameo role in the feature film *Frequency* (2000). Earlier, Greene had appeared in musicals while at Harvard, and while at Cornell, he had a part in

Harold Pinter's play *Betrayal* at a community theater. "It's a release, a way to enter a new world," Greene told Boss about his sporadic ventures into acting. "The things you think about are totally different from what you think of in a normal research day. Issues of human character and genuine human response are at the other end of the universe from trying to figure out why this string vibrates this way or that." Greene is interested in psychology and issues of human consciousness, and at one time he competed in judo. He was the director of the Theoretical Advanced Study Institute at the University of Colorado Boulder, and is on the editorial boards of major publications in theoretical physics. When he was thirty, Greene won a Young Investigator's Award from the National Science Foundation. The next year he won an Alfred P. Sloan Foundation fellowship.

## Further Reading

"A Greene Universe." *Scientific American* 21 Apr. 2000: 36. Print.

Nash, J. Madeleine. "Unfinished Symphony." *TIME Magazine* 31 Dec. 1999: 36. Print.

Senior, Jennifer. "He's Got the World on a String." *New York* 1 Feb. 1999: 33. Print.

# Hansch, Theodor W.

## German physicist

**Born**: October 30, 1941; Heidelberg, Germany

On October 4, 2005, the Royal Swedish Academy of Sciences announced that Theodor Hansch was one of three recipients of the Nobel Prize in Physics. He shared half of the award with the American John L. Hall, while the other half went to American Roy J. Glauber for his work in reconciling quantum mechanics to the behavior of light. In supplementary information accompanying the press release announcing the awards (October 4, 2005), the Royal Swedish Academy of Sciences explained that Hall and Hansch earned the prize "for their development of laser based precision spectroscopy." The honor was not entirely unanticipated. "I had a spark of hope. . . . I expected I would be on the list," Hansch stated soon after the announcement, as quoted by Michael Pohl for the Associated Press (October 4, 2005). "I was speechless, but of course very happy, exuberant. If I had known, I would have put on a tie."

### Early Life and Education

Theodor "Ted" Wolfgang Hansch was born to Karl E. Hansch, a businessman involved in the export of farm equipment, and Marta Kiefer Hansch, a homemaker, on October 30, 1941, in Heidelberg, a scenic city in the state of Baden-Wurttemberg in southwestern Germany. (Hansch's last name is sometimes spelled Haensch or Hansch.) His first memories are of World War II. "I can still see our family huddled together in the basement bomb shelter of our home in Heidelberg listening to the piercing sound of air raid sirens," he recalled in his autobiographical statement posted on the official Nobel Prize website. When the war ended in 1945, Hansch and his family, which included his brother, Julius, and his sister, Lucia, were forced to share a tiny, ground-floor apartment with war-displaced refugees, following the loss of the family estate in Breslau. Around this time he developed, with his father's encouragement, a love of science. "My father . . . kindled my early interest in science. During the first world war, volunteering at a pharmacy, he became interested in medicine and chemistry. In Heidelberg we lived at Bunsenstrasse 10, in the house that had once belonged to the chemist Robert Bunsen. When I was about six years old, I asked my father what Bunsen had done to have a street named after him. On the next day he brought home a Bunsen burner which we connected to the gas stove in the kitchen. With a sprinkle of table salt, the blue flame turned to a bright yellow. My father explained that this is the characteristic color emitted by sodium atoms that are excited in the flame. It was obvious to me that I had to find out more about light and atoms. A little later, my father took me to visit the metallurgical laboratory of the Heinrich Lanz AG in Mannheim, where I was impressed by researchers in white lab coats who allowed me to look into their fancy microscopes. At a time when other boys dreamt about steering steam locomotives, I started to see myself as a future scientist."

Beginning in 1952, Hansch attended the Helmholtz-Gymnasium in Heidelberg, where his interest in science became an all-consuming passion. "Early on, my interest in science dominated my activities outside school," he recalled in his Nobel Prize autobiography. "I eagerly read popular science and science fiction books from the public library until I learned how to check out textbooks from the University library. I also liked doing experiments with my own hands. Intrigued by the world of chemistry, I started to spend my weekly allowance in pharmacies willing to sell substances like fuming nitric acid or white phosphorous to a young boy who stored his growing collection of chemicals in the bedroom of his parents. After an intimidating accident with bomb-making materials, my interests moved from chemistry to physics and electronics."

After completing his secondary education at the Helmhotz-Gymnasium in 1961, Hansch went on to study physics at the University of Heidelberg. Despite his initial interest in nuclear physics, he became increasingly fascinated with lasers, then a relatively recent invention, following a visit to the Institute of Applied Physics at Albert-Uberle-Strasse. He wrote in his autobiography, "Visiting this laboratory I was awed by the sight of a helium neon laser with its glowing discharge tube emitting an intense collimated beam of red laser light that produced an otherworldly speckle pattern. I sensed a large unexplored new world, and I instantly decided to switch fields." Subsequent to this visit, Hansch studied the emerging field of lasers and the effects of saturation, at both at the graduate and undergraduate level. He received the equivalent of a BS in physics in 1966 and graduated summa cum laude with a doctorate in 1969. For the following academic year, he served as an assistant professor at the University of Heidelberg's Institute of Applied Physics. In 1970 he left what was then West Germany for the United States, having accepted a North Atlantic Treaty Organization (NATO) postdoctoral fellowship at Stanford University in Palo Alto, California. Hansch remained at Stanford after completing his fellowship, moving rapidly up the academic ranks, joining the faculty as an associate professor in 1972 and earning a full professorship in 1975.

## Life's Work

Throughout his time in California, Hansch worked closely with the physicist Arthur L. Schawlow, one of the inventors of the laser. After meeting at a summer school at Carberry Tower in Scotland in 1969, the two kept up a correspondence and Schawlow helped persuade Hansch to come to Stanford. During Hansch's first year in Palo Alto, he and Schawlow applied a charge to a liquid sample of organic dye, creating a short-lived laser beam that was remarkable for both its purity and the range of frequencies it could be tuned to emit. In 1972, using one of these dye lasers to analyze the frequencies at which atomic particles release energy, Hansch and Schawlow obtained extraordinarily precise measurements of the structures of various atoms and molecules. These frequencies, perceived as color, are called spectra, which gives the branch of science known as spectroscopy—the study of light—its name. The techniques for laser spectroscopy that Schawlow and Hansch developed brought a new level of precision to the measurement of atomic structures. For example, Hansch and Schawlow worked out a method of laser spectroscopy that corrected for the distortions generated by the Doppler effect, a phenomenon that alters the frequency at which waves from a particular source are observed when the source of the wave and the observer are in motion relative to one another. Using this technique, in 1972, Hansch and his colleagues observed a quality in the spectroscopic profile of hydrogen known as the Balmer line with unprecedented clarity; two years later they used lasers to obtain a significantly finer measurement of the speed of light, one of the universe's fundamental constants. Both of these areas of

> **"At a time when other boys dreamt about steering steam locomotives, I started to see myself as a future scientist."**

research would figure prominently in Hansch's future scholarship. As Hansch wrote in an article for *Optics and Photonics News* (February 2005), "For me, the Balmer line experiment was the beginning of a long quest for ever higher resolution and accuracy in laser spectroscopy of the simple hydrogen atom which permits unique confrontations between experiment and theory." Hansch and Schawlow's spectroscopy research won them the 1973 California Scientists of the Year award from the California Museum of Science and Industry (now the California Science Center) in Los Angeles; and, in 1981, led to a Nobel Prize in Physics for Schawlow, who shared the award with Kai M. Siegbahn and Nicolaas Bloembergen. Following his 1973 award, he received tenure as an associate professor position at Stanford University, but also received offers of full professorships from the University of Heidelburg, Yale University, and Harvard University. He decided to remain at Stanford, where he became a full professor in 1975.

Another fruitful avenue of research for the pair opened in 1974, when they proposed a method for using lasers to cool atoms to just above absolute zero (approximately –460 degrees Fahrenheit or –273 degrees Celsius), the theoretical temperature at which all molecular movement stops. Three other scientists—Steven Chu, Claude Cohen-Tannoudji, and William D. Phillips—won the physics Nobel in 1997 for fully developing the laser cooling technique, and in 2001 another group of physicists—Eric A. Cornell, Wolfgang Ketterle, and Carl Wieman, who was one of Hansch's doctoral students at Stanford during the mid-1970s—shared the prize for using this laser cooling technique to chill atoms to within 170-billionths of a degree of absolute zero, producing the first example of a Bose-Einstein Condensation, a never-before-seen phase of matter. Hansch, in turn, has done groundbreaking research into Bose-Einstein Condensation.

However, not all of Hansch's work at Stanford proved to be as serious or as worthy of a Nobel Prize. Famed for a lighthearted approach to his work, Hansch, in collaboration with Schawlow, conducted tests in the early 1970s on Jell-O, in an attempt to produce "edible lasers," as he called them. The unadulterated gelatin failed as a medium. The results of the study were informally published—"this may well be the first edible laser material," the note on the experiment proclaimed, according to Hansch's article for *Optics and Photonics News*—and, after further explorations by other researchers, the findings were put to use on semiconductor lasers.

In 1986 Hansch returned to West Germany, where he became a professor of experimental physics and laser spectroscopy at Munich's Ludwig Maximilians University and the director of the Max Planck Institute for Quantum Optics in Garching, a suburb of Munich. (Hansch remained a full professor at Stanford until 1988, when he became a consulting professor, a position he continues to retain.) As the head of one of the Max Planck Society's eighty research centers, Hansch contributed to many different research programs. He continued to refine his measurements of atomic particles, focusing particularly on shifts in the energy levels of hydrogen. In 1992 Hansch and other researchers in his Munich laboratories, building on the work of American physicists, developed a method for holding groups of rubidium atoms in two- and three-dimensional traps called lattices. Six years later Hansch and his colleagues, using atoms held together in a Bose-Einstein Condensate, developed a laser that fired whole atoms rather than the tiny packets of electromagnetic energy called photons that normal lasers emit. Though a group of scientists in the United States had managed to make atoms behave in a similar way the year before, the laser-like flow Hansch and his colleagues were able to create formed a continuous and highly coherent stream rather than a series of disconnected pulses.

## The Optical Frequency Comb Synthesizer

Perhaps Hansch's most important achievement came in 1997, when he and a team of researchers in Garching measured the ultraviolet frequency emitted by a hydrogen atom shifting from one energy level to another. Their calculations determined the frequency with a precision 100 times greater than the most accurate previous measurement. Their achievement, according to Bertram Schwarzschild in *Physics Today* (December 1997), was "the most accurate measurement to date of any frequency in the visible or ultraviolet" light ranges. To accomplish this "tour de force," as Schwarzschild described it, Hansch and his colleagues employed a kind of subatomic ruler that has come to be called the optical frequency comb synthesizer, a device that measures the frequency of electromagnetic waves using a "comb" consisting of hundreds of thousands of incredibly quick laser "teeth." Each tooth in the comb (or mark on the subatomic ruler) is spaced at a perfectly uniform distance from the other and emitted in pulses regulated by an atomic clock. First proposed by Hansch in a confidential (and therefore unpublished) paper in March 1997, the frequency comb had roots in still earlier projects. In 1978 Hansch had joined two other Stanford physicists in proposing and testing a considerably less powerful device that aspired to perform essentially the same function. Commenting on the original paper in light of Hansch's 2005 Nobel Prize, one of his corecipients, John L. Hall, stated, as quoted by Kenneth Chang for the *New York Times* (October 5, 2005), that he considered Hansch's initial proposal either "genius prophetic or absolutely absurd," and that the latter was probably more likely. "I didn't see any way to make that happen." The technology of the time certainly made it difficult to see how the process might work. Instead of the ultra-fast pulses of the later comb—lasting a femtosecond, or one billionth of a millionth of a second—the comb proposed in 1978 could register a picosecond, or one millionth of a millionth of a second. As laser technology developed, the idea continued to be tested, with Hansch, Hall, and others making marked progress toward a more precise device during the 1990s. But not only were the measurements made during these experiments less accurate, they also "used impractically complex setups (basketball-court large, multi-million-dollar expensive, and so complex that they never worked for long)," according to Ludwig Bartels, an assistant professor of physical chemistry/chemical physics at the University of California, Riverside. However, several technological advances soon contributed to the laser comb becoming a reality, among them improvements in fiber-optic cable and refinements in laser technology that made the lasers used in the comb intense and stable enough to bear the heavy burden required of them. Once fully developed, Hansch's frequency comb can "fit on a regular desk," Bartels noted—evidence that, as Hansch stated to Gloria B. Lubkin for *Physics Today* (June 2000), "for high-precision measurements, our apparatus gets smaller and cheaper as we make progress."

Though the frequency comb synthesizer was the product of what Hansch, in his Nobel lecture, termed "curiosity-driven" (rather than "goal-driven") research, the technology has a host of practical applications. One seemingly imminent advance that the frequency comb portends is the production of optical clocks that lose less than a picosecond a day, making them possibly hundreds of times more accurate than today's atomic clocks. This degree of precision could in turn lead to greater control over satellite and space-probe tracking technology. The frequency comb could also be used to examine general and special relativity and the fundamental constants of the universe, which some scientists think might be slowly changing as time passes. Experiments Hansch and his colleagues performed in 2004, however, suggested that these changes are either not occurring, or are occurring at a pace even more minute than can currently be measured. Hansch has also suggested to reporters that recent advances in laser technology might lead to three-dimensional holographic films. "This work has opened up a new frontier in optics and optical technology whose final implications we cannot yet guess," Daniel Kleppner, the Lester Wolfe Professor of Physics at the Massachusetts Institute of Technology (MIT), stated, according to Chang. "That sounds like hyperbole, but I think it's true."

In addition to the Nobel Prize, Hansch has been honored with numerous awards over the years. Some of his laurels include the first-ever Otto Hahn Prize for Chemistry and Physics in 2005, which was given jointly by the German Chemical Society, the German Physical

Society, and the City of Frankfurt, Germany. Also in 2005 Hansch garnered the Frederic Ives Medal, the Optical Society of America's most eminent prize. In 2003 he received the Bundesverdienstkreuz, the highest civilian honor bestowed by the German government, and an analogous award from the German state of Bavaria, the Bayerischer Maximiliansorden. He received the Matteucci Medal from the Italian National Academy of Sciences in 2002; the European Physical Society's Quantum Electronics and Optics Prize in 2001; and in 2000 both the Laser Institute of America's Arthur L. Schawlow Award and the Philip Morris Research Prize for his work on the atom laser. In 2009, Hansch received the James Joyce Award from University College Dublin and, in 2010, he won the Bavarian Constitution Medal in gold, in recognition of his "outstanding services to the general public."

Unmarried, Hansch makes his home in Munich. In July 2001 he cofounded the company Menlo Systems to make frequency comb technology commercially available to other researchers. In his offices at the University of Munich, Hansch keeps a host of toys at hand. "Since the frivolous edible laser experiments at Stanford," Hansch wrote in *Optics and Photonics News*, "I refuse to feel guilty about simply enjoying some playtime in the lab."

### Further Reading

"Femtosecond Comb Technique Vastly Simplifies Optical Frequency Measurements." *Physics Today* June 2000: 19. Print.

Hansch, Theodor. "Edible Lasers and Other Delights of the 1970s." *Optics and Photonics News* Feb. 2005: 14. Print.

Schwarzschild, Bertram. "Optical Frequency Measurement is Getting a lot More Precise." *Physics Today* Dec. 1997: 19. Print.

# Hayhoe, Katharine

## American climate scientist

**Born**: 1972; Etobicoke, Canada

Katharine Hayhoe is the director of the Climate Science Center at Texas Tech University, as well as the head of a consulting company called ATMOS Research, which provides climate projections to clients in both the public and private sectors. She is also the coauthor, along with her husband, evangelical pastor Andrew Farley, of *A Climate for Change: Global Warming Facts for Faith-Based Decisions* (2009). Hayhoe, also an evangelical Christian, told Tom Miller for the PBS show *Nova* (April 27, 2011), "There's often a perceived conflict between science and faith. It's a little bit like coming out of the closet, admitting to people that you are a Christian and you are a scientist." Although many religious and political conservatives profess doubt about the reality of global warming, Hayhoe has stated that human-driven climate change is as certain a phenomenon as the law of gravity or Einstein's theory of relativity; she attributes skepticism about the topic to media coverage that misleadingly portrays climate change as a two-sided issue. "The most recent survey I've seen found that almost 98 percent of scientists in related or relevant fields to climate science agreed that human production of heat-trapping gases was the main or most important influence on climate change today, and was responsible for much of the climate change that we've seen over the last 50 years," Hayhoe told Kate Galbraith for the *Texas Tribune* (September 22, 2011).

Hayhoe explained to a journalist for the Christian environmental website Restoring Eden, "I firmly believe there is no conflict at all between faith in an all-powerful God and understanding that humans are radically altering the face of our planet. In fact, they are completely compatible." She continued, "We already know that bad things happen in our world. Poor choices have consequences. We live this out in our own lives, and now we see the same principle at work at the global scale. Climate change is already affecting our planet and its inhabitants. The greatest impacts are already being felt by the poor and the disadvantaged, who lack the resources to adapt. This is true both here in the United States, as well as in developing nations around the world. As Christians, we are called to love God and love others. Recognizing the reality of climate change and reaching out to help our global neighbors is a tangible expression of this love." Hayhoe, who was an important contributor to the Nobel Prize–winning Intergovernmental Panel on Climate Change (IPCC), believes that everyone can help combat global warming by making simple changes in their daily lives. To further that cause, she gives countless interviews and works to raise public awareness, devoting so much of her free time to lecturing that she has been called the "Climate Evangelist."

## Early Life and Education

Hayhoe was born in 1972 in Etobicoke, a municipality within the city limits of Toronto, Ontario. Both of her parents were teachers. Her earliest memory, she has told journalists, is of stretching out on a blanket in the park at age four and learning to find the Andromeda galaxy in the night sky. "As far back as I can remember, my father was teaching me about the world around us," Hayhoe told Tom Miller, "whether it was memorizing all of the birds that we would see in our backyard, or keeping an eye out for all of the rare wild flowers that there are in Ontario, or the giant telescope that we dragged with us on most of our family vacations. But at the same time, from the very beginning, as he taught me about the world, he also taught me that it was the result of a God who created it. And the more I study the world, the more it seems to me that that is the case."

Hayhoe spent many of her childhood summers at a vacation cottage in the Canadian Shield, an area also known as the Laurentian Plateau. When she was nine, her family moved to the South American nation of Colombia, where her parents worked as missionaries and taught in a bilingual school. Hayhoe spent a lot of her time hiking through the Colombian mountains to visit remote villages and examining the unusual flowers and fruits growing alongside the roads. She returned to Ontario to attend the University of Toronto, where she studied physics and astronomy. As an undergraduate student, she worked in the university's research labs, operating a telescope that studied variable stars and testing satellite instruments that measured the earth's atmosphere. She told Kate Galbraith, "It became apparent [to me] fairly quickly that one of the greatest issues related to atmospheric physics was the fact that humans are actually altering the content of our atmosphere, putting more heat-trapping gases into our atmosphere than would normally be there, and those heat-trapping gases are affecting the climate of our planet." After earning her BS degree in 1994, Hayhoe spent a year backpacking through Europe. She then pursued her graduate degree in atmospheric science under the mentorship of Donald Wuebbles at the University of Illinois at Urbana-Champaign, choosing the school, in part, because it hosted a large and active chapter of the Intervarsity Christian Fellowship. She soon became a leader in the group, and through its activities she met her future husband, Andrew Farley. Hayhoe graduated with an MS degree in atmospheric sciences in 1997 and married Farley in 2000. (Later, in 2010, she received her PhD degree from the school.)

## Life's Work

By 2004 Hayhoe had begun region-specific impact studies with AT-MOS Research & Consulting in South Bend, Indiana, while Farley taught at the University of Notre Dame. That year she appeared as a guest on the National Public Radio program *Talk of the Nation* (August 27, 2004), discussing a paper she had authored for the *Proceedings of the National Academy of Sciences* (*PNAS*) that described how climate change might affect the people, industries, and ecology of California. "We chose California because it is a challenging case," she told the show's host, Ira Flatow. Hayhoe and her team had envisioned two hypothetical scenarios for the study. In the first, the world does not alter its current behavior and continues to liberally use fossil fuels. In the second scenario, scientists look for alternate sources of energy but do not, as Hayhoe noted, take "specific actions to reduce . . . emissions of greenhouse gases." She discovered that neither scenario yielded particularly optimistic results. "We found the impacts [of continued fossil fuel emissions] were widespread over almost every sector that we looked at," she told Flatow. "We found that with higher temperatures, you get more frequent and severe periods of extreme heat. As you can imagine, that affects human health in both the urban and rural areas. We found reductions in snowpack in the mountains, impacts on water supply in California, impaired quality of wine grapes. . . . [There would be major damage to] the vegetation and the ecosystems that

> **"If we wait more than 10 or 20 years before taking action on climate change, extreme measures could be required to prevent what scientists call 'dangerous human-driven change.'"**

make California such a unique place." She continued, "The Central Valley might feel more like Death Valley." In response to skeptics, including the National Center for Policy Analysis, which described Hayhoe's study as "a doomsday prediction by climate alarmists," Hayhoe defended her figures. "From the science side, I would say that our study is actually rather conservative," she told Flatow. "The whole point of the study is to show that we have the potential to determine the outcome of our own future. The choices that we make today are going to have huge impacts on the amount of change that we see in the future."

Hayhoe has conducted similar region-specific studies in other parts of the country. Her mid-range climate predictions indicate that summers in West Virginia will soon feel more like those in Arkansas or Kansas. "The West Virginia that people are used to, that they grew up in, that their parents grew up in, it's not going to be the same place their children grow up in," she told Pam Kasey for the (West Virginia) *State Journal* (February 2007). "We are seeing all kinds of signatures of [climate] change [in West Virginia and] all over the country. For example, trees and flowers are blooming earlier in the year, and we're seeing new species growing where they didn't grow before because it was too cold." By the end of the century, she predicts that Cincinnati, Ohio—which now gets 18 days per year of temperatures over

90 degrees—could be subject to 85 days of such extreme heat. "In addition," as Bob Downing reported for the *Akron (Ohio) Beacon Journal* (July 29, 2009), "Ohio will get more heavy rains that would trigger flooding; winters would be shorter; smog would continue to be a problem in Northeast Ohio and Cincinnati; and the water level on Lake Erie will fall about 18 inches. Ohio farmers will also be heavily impacted. Crops and livestock will face substantially more heat stress, decreasing crop yields and livestock productivity. The growing season will become up to six weeks longer but crop pests like the corn earworm are likely to spread."

Hayhoe has also addressed the recent outbreak of extreme weather in Texas, which experienced a crippling drought and record-breaking temperatures in 2011. She told Galbraith, "We have already altered the background conditions of our atmosphere, through increasing our production of these heat-trapping gases and increasing the average temperature of the atmosphere. . . . We have changed the background conditions in terms of not just the temperature but the humidity and also the weather patterns that we've experienced. So in that sense, every event that happens—snowstorm, heat wave, drought, flood—every event that happens has some contribution or component of climate change in it, because we've changed the background conditions."

Citing the drought in Texas and other such extreme events that have occurred over the past decade, Hayhoe has deemed strange weather patterns to be the products of "global weirding," a concept broader than simple global warming, which reflects only one aspect of climate change. "It's certainly true that we have an enormous diversity of weather on our planet and that's why we have to be careful not to immediately jump to conclusions about what could be causing a specific weather event," Hayhoe told Daniela Minicucci for the Canwest News Service (May 26, 2011). She used an analogy to describe the new and disturbing weather trends. "When we throw a pair of dice, we always have a chance of rolling two sixes," she told Minicucci. "That's the natural variability. What climate change is doing, however, is slowly, one by one, removing one other number at a time and replacing it with an identical six. So our chances of rolling double sixes for many extreme events have slowly been increasing."

Hayhoe began teaching at Texas Tech University in 2005, and she became director of the Climate Science Center at the university in 2011. In 2007, as an associate professor in the Department of Geosciences, she contributed to the United Nations Intergovernmental Panel on Climate Change (IPCC). The study included scientists from 130 countries who reached a nearly unanimous conclusion about human-driven climate change. "The further we go the more evidence we see that our predictions are right," Hayhoe told Caleb Hooper for the *Texas Tech Daily Toreador* (February 16, 2007). "If we wait more than 10 or 20 years before taking action on climate change, extreme measures could be required to prevent what scientists call 'dangerous human-driven change.'" The IPCC shared the 2007 Nobel Peace Prize with former US vice president and environmental activist Al Gore "for their efforts to build up and disseminate greater knowledge about man-made climate change, and to lay the foundations for the measures that are needed to counteract such change," as noted on the Nobel Foundation's website.

## Family and Faith

In 2009 Hayhoe and her husband, Andrew Farley, co-authored *A Climate for Change: Global Warming Facts for Faith-Based Decisions*. Farley is a pastor at Ecclesia Church in Lubbock, Texas, as well as a tenured professor of linguistics at Texas Tech. The two decided to write the book because they were concerned that evangelical Christians were being misinformed about climate change. (While other religious groups have called for action to prevent climate change, evangelicals remain divided.) "This book came from conversations with friends who wanted to know the truth about climate change from a fellow Christian—someone they could trust," Hayhoe stated, as quoted in a Texas Tech press release (October 29, 2009). "We realized we had an incredible opportunity to speak out on one of the most pressing issues facing our generation. Most Christians are not scientists, and it's hard to say how many scientists are Christians. In our family, we have both."

Hayhoe and Farley stress the importance of separating science and politics. "Many Christians may feel like they've been burned by scientists in the past, perhaps on issues related to the sanctity of life, or the creation-evolution debate. Complicating the issue is the fact that climate change has become so politicized, albeit falsely so. The result is that conservatives, and many Christians, tend to view climate change as yet another environmental issue driven by a liberal agenda with ulterior motives," Hayhoe told the writer for the Restoring Eden website. "The science of climate change has nothing to do with red politics, or blue politics, or any kind of politics. It's a simple matter of temperature readings and long-term trends that have been happening over the last few centuries."

Hayhoe recalled that Farley had once been a "climate change skeptic" because of the misinformation perpetuated by conservative pundits. They spent many late nights together, she said, verifying each piece of evidence in support of climate change until he was convinced. The couple hopes to convince others with their book, which addresses the issue from both a scientific and theological standpoint. Hayhoe and Farley conclude with a biblical call to action. "As we say in the book," Hayhoe told Jim DiPeso for the Republicans for Environmental Protection website, "God may have given us the ostrich, but it wasn't as a mythical example to imitate when confronted with unpleasant facts we'd rather ignore. Instead of burying our heads in the sand, we should take Gideon as our model, who examined all possible options and then when he'd had his questions answered, took action."

Hayhoe served as the lead author on the United States Global Change Research Program report "Global Climate Change Impacts in the United States" (2009) and the National Research Council report "Climate Stabilization Targets: Emissions, Concentrations, and Impacts Over Decades to Millennia" (2010). She recently won funding from the United States Geological Survey (USGS) to develop a national database of climate-change projections and is a principal investigator with the Department of the Interior's South-Central Climate Science Center. Hayhoe lives in Lubbock, Texas, with her family.

## Further Reading

Downing, Bob. "Scorching Summers a Definite Possibility." *Ohio.com*. Akron Beacon Journal, 29 July 2009. Web. 19 Dec. 2011.

"Dr. Katharine Hayhoe Talks About California's Climate Change." *Talk of the Nation*. NPR. 27 Aug. 2004. Radio.

Hayhoe, Katharine. "Thermometers Don't Lie: An Interview with Author Katharine Hayhoe." *Restoring Eden*. Restoring Eden, 2010. Web. 19 Dec. 2011.

Kasey, Pam. "Atmospheric Scientist Offers View of a Changed West Virginia." *State Journal* [West Virginia] 9 Feb. 2007: 6. Print.

"Katharine Hayhoe." *The Secret Life of Scientists and Engineers* (*Nova* web video series). PBS, 2011. Web. 19 Dec. 2011.

Minicucci, Daniela. "Spree of Deadly Storms Points to 'Global Weirding.'" *Global Toronto*. Shaw Media, 26 May 2011. Web. 19 Dec. 2011.

# Herr, Hugh

## American biophysicist and engineer

**Born**: October 25, 1964; Lancaster, Pennsylvania

The biophysicist and mechanical engineer Hugh Herr was a highly skilled, experienced mountaineer when, at age seventeen, he got lost during a storm while climbing with a friend on New Hampshire's Mount Washington. By the time the two were rescued, their long exposure to frigid temperatures had caused severe frostbite—in Herr's case, in his lower legs. The tissue damage proved to be irreparable, and Herr's legs were amputated below his knees. More agonizing to him than the extreme pain he suffered physically for many months was the knowledge that a volunteer rescuer, Albert Dow, had died in an avalanche while searching for him on Mount Washington. His anger at himself for his mistakes on the mountain and his shame regarding Dow's fate inspired in him an "extraordinary swelling of energy and a desire to not wallow in self-pity but to do something worthwhile with my life," he told Terry Gross for the National Public Radio program *Fresh Air* (August 10, 2011). That energy and desire led Herr, who had valued climbing over school, to become a fiercely dedicated undergraduate and graduate student. He went on to become a leading researcher in the biological and physical aspects of human movement and a pioneering designer of state-of-the-art artificial knees, ankles, and limbs for amputees and aids for people with limited mobility. Herr told Logan Ward for *Popular Mechanics* (September 29, 2005) that his work focuses on the "difficult problem of using modern synthetic materials to replace the extraordinary systems nature has given us." He told Andy Greenberg for *Forbes* magazine (November 29, 2009), "I feel a responsibility to use my intellect and resources to do as much as I can to help people. That's Albert Dow's legacy for me."

Herr is the director and principal investigator of the Biomechatronics Research Group at the Media Lab of the Massachusetts Institute of Technology (MIT), where he is an associate professor in the

Department of Media Arts and Sciences and the MIT-Harvard Division of Health Sciences and Technology. Biomechatronics is a field in which biology, mechanics, and electronics intersect. Herr and his team have created devices that have benefited amputees and others both physically and emotionally, helping them to "recover not just their bodily movements but the quality of life they enjoyed before their need for prostheses," Herr said in a press release distributed by PR Newswire (September 1, 2011). People who have had legs amputated because of accidents or war-related injuries constitute only a small percentage of amputees in the United States; far more have lost legs because of diabetes. In part because of the growing incidence of obesity-triggered diabetes, the number of amputations is expected to increase greatly.

The bionic knees, ankles, and limbs invented in Herr's lab embody major advances in the comfort, durability, energy efficiency, safety, flexibility, adaptability, mobility, and even aesthetics of prosthetics. "We want the bionic limb to have a humanlike shape, but we don't want the bionic limb to look human," Herr told Gross. "We want it to look like a beautiful machine, to express machine beauty as opposed to human beauty. . . . We don't want the user of these prosthetic limbs to be ashamed of their body, ashamed of the fact that part of their body is artificial. We want them to celebrate it. So to do that, we need to put forth synthetic structures that are elegant." Herr is the founder and chief technology officer of iWalk, which markets inventions such as the PowerFoot motorized ankle-foot prosthesis.

"Below the knee . . . down to the floor, I'm artificial," Herr told Gross. "I'm titanium, carbon, silicon, a bunch of nuts and bolts. My limbs that I wear have five computers, twelve sensors, and muscle-like actuator systems that enable me to move throughout my day." Herr owns many pairs of limbs, with feet designed for walking, jogging, running, and climbing on different surfaces. His climbing prostheses have enabled him to climb more efficiently than he did with his own legs. "In the very near future," he said in the press release, "as we continue to learn how to efficiently connect prostheses to the human body, both electrically and mechanically, we will expand controllability and feeling of the artificial limb."

Herr's honors include the Boy Scouts of America's Young American Award in 1990, the Next Wave: Best of 2003 Award from *Science Magazine*, the 2005 Breakthrough Leadership Award from *Popular Mechanics*, the 2007 Heinz Award in Technology, the Economy and Employment, and the Spirit of da Vinci Award from the National Multiple Sclerosis Society's Michigan chapter in 2008.

## Early Life and Education

The youngest child of John and Martha Herr, Hugh Miller Herr was born in Lancaster, Pennsylvania, on October 25, 1964. On his father's side, Herr is descended from a Mennonite minister who immigrated to Lancaster from Switzerland in 1710 to escape religious persecution. He grew up on the family farm in Lancaster with his siblings, Hans, Beth, Ellen, and Tony. The Herr siblings assisted with farm chores from an early age and helped their father with his work building houses. In his acceptance speech for the Heinz Award, Herr said that his parents had impressed upon him "the simple idea that there's no obstacle too great when confronted with the power of the human spirit."

During the summers the Herrs traveled west to mountainous areas of the United States and Canada to camp and hike. Herr's brothers Tony and Hans became avid climbers, and by age seven, Herr was accompanying them. He became particularly interested in mountaineering, which encompasses activities including rock climbing, ice climbing, and bouldering, each with or without ropes or other specialized equipment. Succeeding at difficult climbs requires an enormous amount of knowledge about the environment, safety hazards and precautions, proper use of equipment and techniques, and human physiology.

By age seventeen Herr had become well known as one of the boldest and most skilled young climbers on the East Coast, having conquered a number of the most difficult climbs at a younger age than anyone else. Little interested in academic subjects, he had enrolled in a vocational program at Penn Manor High School, a local public school. He worked mostly in the machine shop, where he learned to make things with metal and wood.

In January 1981 Herr and an older friend from Lancaster, Jeff Batzer, climbed Pinnacle Gully in Huntington Ravine on Mount Washington.

Rising 6,288 feet, the mountain is the highest peak east of the Mississippi River. According to the Mount Washington Observatory website, its climate is akin to "that of Northern Labrador, hundreds of miles further north. Three major storm tracks converge over the mountain, forming harsh and turbulent weather conditions." Furthermore, its "combination of the extreme wind, fog, wet, and cold" has led Mount Washington to be called Home of World's Worst Weather. Describing the trail Herr and Batzer took, the observatory website states, "During the snow-free months the Huntington Ravine Trail is regarded as the most exposed and intimidating hiking path in the White Mountains. Add a little bit of snow and making your way through the Ravine automatically becomes a full-blown mountaineering challenge."

In January 1982 Herr, then a junior in high school, and Batzer returned to Mount Washington. Although it was snowing the observatory had detected no danger of avalanches in Odell Gully, a seven-hundred-foot ice face in the Huntington Ravine, so Herr and Batzer decided to climb there. Warmly dressed and well equipped, they started out on the morning of Saturday January 23, 1982. Soon, to lighten their load, they abandoned their sleeping bag and other camping provisions. When they reached the top of the gully, they decided to continue on to the mountain's summit. Before long, blowing snow caused visibility to plummet to less than thirty feet, and gusts of wind buffeted them. With an inadequate map and no compass, Herr and Batzer became lost. During the next three days and nights, they became increasingly disoriented and incapacitated by the cold, suffering from frostbite, exhaustion, and thirst.

The failure of Herr and Batzer to return on Saturday to the climbers' facility where they had stayed Friday night triggered the formation of search teams by the Appalachian Mountain Club (AMC), the Mountain Rescue Service, the New Hampshire Fish and Game Department, and the US Forest Service. On Sunday morning employees and volunteers began to look for them, knowing only that Herr and Batzer had intended to climb Odell Gully, not that they had extended their climb. The teams searched for three days, despite inclement weather and the death of one volunteer rescuer, twenty-eight-year-old Albert Dow, in an avalanche. On the afternoon of January 26, Melissa "Cam"

Bradshaw, an AMC member, found Herr and Batzer by chance while snowshoeing. Bradshaw had no radio or other means of communication; neither did a pair of hikers she came upon and directed to the men. Several hours passed before Herr and Batzer were taken by helicopter to a nearby hospital. Herr was later transferred to a Philadelphia medical center. There, in March, after repeated efforts to save Herr's legs failed, both legs were amputated below the knee. "I was told by my doctor that I would never climb again," he told Frank Moss, the author of a book about MIT's Media Lab, *The Sorcerers and Their Apprentices* (2011).

After leaving the hospital Herr used a wheelchair for a month before he was fitted for his first pair of artificial limbs. Made of plaster of paris, they were heavy, clumsy, and uncomfortable and designed for very slow walking on horizontal surfaces, not for rapid walking, running, or rock climbing. Nevertheless, by the summer of 1982, Herr had resumed climbing. Having completed his junior year of high school with home tutors, he returned to Penn Manor. He soon began to design his own prostheses in the school's machine shop, using metals and acrylic resin laminates. "Building and testing every prosthetic model he dreamed up, he became his own guinea pig, as unafraid of pain and failure as he was unbothered by the countless falls he took as a climber," Eric Adelson wrote for *Boston Magazine* (March 2009). Herr told Charles J. Murray for *Design News* (December 2010), "I realized that the artificial part of my body is a blank slate from which to create. The limitations were really limitations of technology. My biological body was not disabled. My artificial limbs were disabled." In the spring of 1983, a feature article about him appeared in *Outside* magazine. The cover photo showed him sitting on a boulder, equipped with prosthetic legs and seated next to feet of his own design.

Herr graduated from high school in 1983. During the next two years, he spent much of his time climbing. In 1985 he enrolled at Millersville University, a state university near Lancaster. He immediately discovered that he loved his courses, particularly mathematics; he eventually chose to major in physics. He excelled academically, later telling Adelson that his "intellectual birth was in college." Herr also worked with Barry Gosthnian, a prostheticist in Mechanicsburg,

Pennsylvania, with whom he collaborated on a design for a socket that cushions an amputee's stump at the points of contact with the prosthesis; theirs incorporated fluid-filled bladders. In May 1990, just before Herr earned a bachelor's degree in physics, he and Gosthnian were awarded a patent, the first of many for Herr, for their "inflatable

> **"In the very near future, as we continue to learn how to efficiently connect prostheses to the human body, both electrically and mechanically, we will expand controllability and feeling of the artificial limb."**

limb prosthesis with preformed inner surface." In 1993 Herr earned a master's degree in mechanical engineering from MIT. He completed a PhD in biophysics from Harvard University in 1998. His dissertation was entitled "A Model of Mammalian Quadrupedal Running."

## Research and Invention

Herr has worked in MIT's Media Lab since 1998, beginning as a post-doctoral fellow. He taught at Harvard Medical School, as a member of the Department of Physical Medicine and Rehabilitation, from 1999 to 2004; he became a professor at MIT in 2004. He has supervised the preparation of many undergraduate and graduate theses and doctoral dissertations. As the head of the Biomechatronics Group, he guides research aimed at understanding the neurological, muscular, and other elements involved in balance and mobility. The lab then uses such knowledge to create devices that mimic the body's musculoskeletal design, muscle behavior, range of motion, and mechanisms that enable it to respond instantaneously and continuously to such external conditions as a bumpy sidewalk.

Devices designed in Herr's lab include a microprocessor-controlled prosthetic knee—formally, the variable-damper knee prosthesis—that "automatically adapts to an individual's walking style and environment," according to Össur, the Icelandic-based company that markets it as the Rheo Knee. In 2001 Herr unveiled a robotic fish that swam by means of electronically activated frog muscles. A "low-cost, body orientation sensor" followed three years later. According to Charles J. Murray, the lab's PowerFoot prosthesis enables below-knee amputees "to climb stairs, traverse ramps, walk fast, and exert a level of force that's comparable to that of a biological ankle. And it does it all while enabling users to walk with a normal gait." The PowerFoot has "a small [direct-current] motor, a transmission, and a series spring," Murray explained. "Together, the three elements work with a carbon composite spring foot to provide the power that would otherwise be missing from an amputee's step." *Time Magazine* ranked the Rheo Knee and the PowerFoot among the best inventions of 2004 and 2007, respectively.

Herr envisions that within the foreseeable future, "the artificial prosthesis will become more intimate with the biological human body," as he told Terry Gross. One way of accomplishing this will involve attaching a prosthesis "mechanically, by a titanium shaft that goes right into the residual bone, where you can't take the artificial limb off," he said. An electrical interface between the body and prosthesis may provide another means for this; as Herr explained, "the nervous system of the human will be able to communicate directly with the synthetic nervous system on the artificial limb."

Herr and his team have also designed an ankle-foot orthosis for people with intact limbs who suffer from foot-drop gait, a consequence of certain types of strokes. To help those whose legs are paralyzed or whose leg muscles are impaired, the team is working on exoskeletal robotic structures. "That's a robot that wraps around a biological limb," Herr explained to Gross. The robot will push on "the impaired biological limb in just the right way to allow a person to stand, walk, and even run." He and his colleagues are also constructing "exoskeletal structures that [augment] human capability beyond what nature intended." In the future Herr envisions, "when a person, for example, goes jogging, they'll routinely wear robots . . . to protect their joints."

Herr is optimistic about his own future as well. He told Gross, "My biological body will degrade in time due to normal, age-related degeneration. But the artificial part of my body improves in time because I can upgrade. I can get the best computer, the best motor system, the best artificial intelligence." He continued, "So I predict that when I'm eighty years old, I'll be able to walk with less energy than is required of a person that has biological legs. I'll be more stable. I'll probably be able to run faster than a person with biological limbs. . . . In fact the artificial part of my body is in some sense immortal."

## Personal Life

Herr and his wife, Patricia Ellis Herr, have two daughters, Alexandra and Sage, both of whom are veteran hikers and climbers. Patricia Herr is the author of *Up: A Mother and Daughter's Peakbagging Adventure* (2012).

## Further Reading

Adelson, Eric. "Best Foot Forward." *Boston.* Metrocorp, Mar. 2009. Web. 27 Jan. 2012.

"The Double Amputee Who Designs Better Limbs." *Fresh Air.* Natl. Public Radio, 10 Aug. 2011. Web. 26 Jan. 2012.

Greenberg, Andy. "A Step beyond Human." *Forbes.* Forbes.com, 25 Nov. 2009. Web. 25 Jan. 2012.

"Hugh Herr." *Biomechatronics Group.* Massachusetts Institute of Technology, n.d. Web. 26 Jan. 2012.

McCarthy, Alice. "Hugh Herr: Back on Top." *Science Careers.* Amer. Assn. for the Advancement of Science, 20 June 2003. Web. 29 Jan. 2012.

Moss, Frank. The Sorcerers and Their Apprentices: How the Digital Magicians of the MIT Media Lab Are Creating the Innovative Technologies That Will Transform Our Lives. New York: Crown, 2011. Print.

Murray, Charles J. "Bionic Engineer." *Design News* Dec. 2010: 38–41. Print.

Osius, Alison. *Second Ascent: The Story of Hugh Herr.* Harrisburg: Stackpole, 1991. Print.

Ward, Logan. "2005 Popular Mechanics Breakthrough Awards." *Popular Mechanics.* Hearst Communication, 29 Sep. 2005. Web. 26 Jan. 2012.

# Heuer, Rolf-Dieter

German particle physicist and director-general of CERN

**Born**: 1948; Boll, Germany

For most of his life, Rolf-Dieter Heuer has dedicated himself to advancing the field of particle physics. Heuer once said, "I was always motivated to work at the energy frontier—wherever that was," as quoted by Virginia Gewin in *Nature* (January 2008). Heuer spent fourteen years with the European Organization for Nuclear Research (CERN), a renowned particle-physics laboratory on the Swiss-French border, where he climbed through the ranks to serve as spokesperson of a project involving the design and construction of a detector for the lab's flagship accelerator. After leaving CERN in the late 1990s, Heuer returned to his native Germany after accepting a faculty position at the University of Hamburg. There Heuer was among a group of researchers responsible for gathering data for an experiment involving the largest particle accelerator at Deutsches Elektronen-Synchrotron (DESY), the country's foremost research facility for high-energy particle physics. By the early 2000s he had joined the staff at DESY, serving as its director of research for particle physics. Heuer's career came full circle in 2009 when he was appointed director-general of CERN. Since taking the reins he has overseen the repair of the Large Hadron Collider (LHC) and the ongoing search for new phenomena such as the Higgs boson.

### Early Life and Education

Rolf-Dieter Heuer was born in 1948 in the municipality of Boll, which is located in the district of Göppingen in Baden-Württemberg, Germany. His father worked as a clerk, and his mother was a homemaker.

In 1969 Heuer attended the University of Stuttgart, where he studied nuclear physics before earning his degree in 1974. He then pursued his doctoral studies at the University of Heidelberg, where he was supervised by Professor Joachim Heintze.

## Early Research

In 1977, after earning his PhD, Heuer accepted a faculty position as a research scientist at the University of Heidelberg. There Heuer immediately became involved with the approved JADE experiment, which was a scientific collaboration between participating universities from three countries—Japan, Deutschland (Germany), and England—lending their names to the acronym. Heuer was responsible for managing the building of the particle detector, also called JADE, whose research was subsequently performed at the Positron-Electron Tandem Ring Accelerator (PETRA) housed at the Hamburg-headquarters of DESY, Germany's biggest research center for particle physics. (An accelerator stimulates subatomic particles to a high velocity and then initiates a high-impact collision with other particles in order to gain a better understanding of the structure of the universe.)

During his five-year tenure at the University of Heidelberg, Heuer narrowly avoided a maximum critical accident (MCA) following the breakdown of one of the conductors at DESY just as an experiment was about to be conducted, resulting in dangerous, high-risk conditions. He told Anna-Cathrin Loll for the *Asia Pacific Times* (September 2009): "This made me understand the importance of continuous quality controls."

## Work at CERN

In January 1984 Heuer left Heidelberg and moved to Geneva, Switzerland, where he joined the staff of CERN, the largest particle-physics lab in the world. Heuer was entrusted with the high-risk responsibility of supervising the design and construction of another particle detector during the experimental phase. The detector in question was the Omni-Purpose Apparatus for the LEP (more commonly known as OPAL), one of four particle detectors constructed at CERN; the three others were ALEPH, DELPHI, and L3. "If only one of the conductors rips, the whole thing won't work anymore," he said in his interview with *Asia Pacific Times*. "With 20,000 conductors, you can imagine that you do lose some sleep over it."

When the biggest electron-positron accelerator, the Large Electron-Positron (LEP), first entered the startup phase in 1989, Heuer served as the run coordinator, a role he held for three years. According to the CERN website, the Large Electron Positron collider "was the accelerator that put the Standard Model of particle physics through its paces. The numerous data taken by the four experiments well and truly tested the model to an incredible level of precision. The experimental results agreed with the theoretical predictions and helped to establish the Standard Model's validity."

In 1994 Heuer was named spokesperson for OPAL, which was among the more renowned investigations in particle physics, and he was charged with overseeing everything that involved the collaborative effort, including a staff of more than three hundred physicists. His duties also included analyzing the data from LEP1 and the energy-upgrade program of LEP1, which converted it into a new accelerator named LEP2. Heuer remained in the post as OPAL's spokesperson until August 1998, leaving after four years because he no longer felt challenged.

### Research in Germany

That same year Heuer returned to Germany and joined the faculty of the University of Hamburg, where he held a professorship. During this period Heuer was among a group of scientists conducting research on the H1 experiment at the Hadron Electron Ring Accelerator (HERA). According to the website for DESY, "HERA was the only storage ring facility in the world in which two different types of particle were accelerated and then collided head-on. This concept required two different accelerators . . . in which the two kinds of particle were accelerated separately and then brought to collision at highest energies using sophisticated beam guidance systems. No one had ever tried to build such a facility before." Another notable achievement was Heuer's establishment of Forschung mit Leptoncollidern (Research with Lepton Colliders), a group comprised of young physics students from German universities.

In December 2004 Heuer was hired by DESY to work for their lab as a director of research for their national particle-physics program. (The DESY lab was a member of the Helmholtz Association of German Research Centres, Germany's biggest scientific institution.)

In an effort to bolster DESY's standing as Germany's main lab for particle physics and reinforce its ties to CERN, Heuer participated in two experiments, ATLAS and CMS. Both were conducted at the most powerful particle accelerator in the world, the Large Hadron Collider (LHC), whose experiments, according to the LHC website, "are expected to address questions such as what gives matter its mass, why nature prefers matter to anti-matter, and how matter evolved from the

> **"I was always motivated to work at the energy frontier—wherever that was."**

first instants of the universe's existence." Following is a description of both experiments according to the LHC website: "Designed to see a wide range of particles and phenomena produced in LHC collisions, each involves approximately 2,000 physicists from some 35 countries. These scientists use the data collected from the complex ATLAS and CMS detectors to search for new phenomena. . . . They also measure the properties of previously-discovered quarks [matter particles] and bosons [force-carrying particles] with unprecedented precision, and are on the lookout for completely new, unpredicted phenomena." Additionally, Heuer was responsible for fostering the lab's involvement in research and development for the International Linear Collider (ILC), an electron-positron collider that has yet to be built, as well as upgrading the luminosity for the LHC.

## Return to CERN

Heuer's profile in the world of particle physics was considerably heightened in December 2007 when he was selected to replace Robert Aymar as director-general of CERN. Following the end of Aymar's tenure in 2008, Heuer took over the five-year post on January 1, 2009, in the midst of a worldwide financial crisis. He also inherited another immediate challenge: The previous September, the LHC—which had ended its experimental phase and was weeks away from its highly anticipated startup on October 21, 2008—experienced an electrical issue. In an article for *Nature* (February 23, 2010), Geoff Brumfiel explained, "A connection between two superconducting cables developed a small amount of resistance, which warmed the connection until the cables—cooled by liquid helium to superconducting temperatures—lost their ability to carry current. Thousands of amps arced through the machine, blowing a hole in its side and releasing several [tons] of liquid helium. The expanding helium gas created havoc, spewing soot into the machine's ultraclean beamline and ripping magnets from their stands." As a result, the LHC was forced to shut down while extensive and necessary repairs were being made.

With increasing media interest in the LHC, CERN was thrust into the global spotlight. The lab was featured in a scene in producer Ron Howard's 2009 thriller, *Angels and Demons*, starring Tom Hanks. Heuer sought to capitalize on this attention by repositioning CERN as more than just a European particle-physics laboratory, but rather as a renowned international hub for scientific research.

A few months into his term as director-general, Heuer promoted a corporate culture that encouraged open communication and sought to bridge the gap between management and employees. To that end Heuer kept the staff informed, communicating major news via e-mail messages and including regular updates regarding the repairs of the LHC in the company's weekly bulletin; he also became more of a fixture at the CERN lab.

Heuer became equally concerned about establishing and maintaining clear and open lines of communication with laboratories, institutes, and governments around the world. Those external parties included the governments of CERN's twenty member countries, who are responsible for financing its programs and managing activities at its facilities, as well as other nonmember nations with whom CERN has cooperation agreements to use the CERN lab for experiments and research. For this reason Heuer created an external relations office to serve as liaison. With the express goal of promoting the open exchange of ideas and information, Heuer launched the CERN Global Network; he also began traveling around the world and giving lectures about particle physics. Another project involved establishing a new center at CERN to analyze and interpret the data output from the LHC.

## Setting Records

Once again all eyes were on CERN on November 20, 2009, with the successful restarting of the LHC, and three days later the data from the first collisions were recorded in all four of its detectors—ATLAS, CMS, ALICE, and LHCb. On November 30, 2009, the LHC set a new record; it became the highest-energy particle accelerator in the world by circulating protons to 1.18 TeV (trillion electron volts), surpassing the previous eight-year record of 0.98 TeV set by the Tevatron, a particle accelerator located at the Fermi National Accelerator Laboratory outside of Batavia, Illinois. In early December the LHC set another world record when the intensity of the beam was increased and the high-impact collision between the two proton beams was documented at 2.36 TeV. Following a shutdown of the LHC in mid-December, the accelerator started up again in February, operating at less than half of its full capacity. Yet another record was set in March 2010 when collisions between two proton beams were recorded at 3.5 TeV per beam (a total energy level of 7 TeV); at the time this was the highest level of energy that had been achieved in a particle accelerator. This event, which was covered by more than a hundred journalists, marked the official start of the LHC research program. Although no discoveries had yet been made with the LHC, Heuer expressed hope that the LHC would be able to locate dark matter, which is invisible and accounts for about 90 percent of the mass of the universe.

In July 2010 CERN found itself in the spotlight again when initial results from the LHC were presented at the thirty-fifth International Conference on High-Energy Physics, which was attended by French president Nicolas Sarkozy. In September 2010 scientists at CERN observing CMS, one of the detectors of the LHC, reported that they had detected a foreign, ridge-like formation produced by the proton collisions. Although these scientists have not yet been able to determine what it is, they published their findings in an effort to promote discussion within the scientific community. CERN scientists continued to observe proton-proton collisions with the CMS detector until the end of October 2010. This was followed by a switch from protons to lead ions for the rest of the year in an effort to more closely mimic conditions similar to the Big Bang.

In 2011 Heuer's staff reported observing what they believed to be Higgs boson, a theoretical, invisible particle (sometimes referred to as the "God particle") at two of the LHC's detectors, ATLAS and CMS. The Higgs boson is thought to give mass to everything in the universe. "It is responsible for the mass of fundamental particles," Michael Barnett, senior physicist at the Lawrence Berkeley National Laboratory and coordinator of education and outreach at CERN, told Jenny Marder for the blog *Rundown News* (December 13, 2011). He explained, "Without that, you don't get the stars and the planets and the universe that we see today. In that sense, it's responsible for our existence. . . . Without it, you'd have this cold, dark universe." Despite the lack of definitive evidence, Heuer is encouraged by the findings and believes that his lab will have an answer regarding the existence of the Higgs boson at some point.

## Personal Life

Heuer lives in the Pays de Gex, France, with his wife, Brigitte. He provided the foreword to the 2009 book *LEP: The Lord of the Collider Rings at CERN, 1980–2000*, written by Herwig Schopper.

## Further Reading

Brumfiel, Geoff. "Did Design Flaws Doom the LHC?" *Nature*. Nature Publishing Group, 23 Feb. 2010. Web. 17 Jan. 2012.

Cookson, Clive. "Domestic Science." *Financial Times*. Financial Times, 23 Sep. 2011. Web. 17 Jan. 2011.

Gewin, Virginia. "Movers: Rolf-Dieter Heuer, Director-General, CERN, Geneva, Switzerland." *Nature* 451.7178 (2008): 602. Web. 17 Jan. 2012.

Loll, Anna-Cathrin. "Lord of the Particles." *Asia Pacific Times*. Asia Pacific Times, Sep. 2009. Web. 18 Jan. 2012.

Marder, Jenny. "Hunt for Higgs Continues; Scientists Work to Separate the 'Signal from the Noise.'" *The Rundown*. MacNeil/Lehrer Productions, 13 Dec. 2011. Web. 18 Jan. 2012.

Sopova, Jasmina. "UNESCO and CERN: Like Hooked Atoms." *UNESCO Courier* (Jan.–Mar. 2011): 48–49. Web. 17 Jan. 2011.

Yurkewicz, Katie. "A New Leader for CERN." *Symmetry Magazine* 6.2 (2009): n. pag. Web. 17 Jan. 2011.

# Higgs, Peter

### British physicist

**Born**: May 29, 1929; Newcastle upon Tyne, England

As the Large Hadron Collider (LHC)—the world's largest and most powerful particle accelerator, or atom smasher—neared start-up in September 2008, the theories of the British physicist Peter Higgs drew a great deal of media attention. That is because the primary goal of those who designed the collider and the more than one hundred nations that paid for its construction is to produce evidence that Higgs's theories—actually, hypotheses—are correct. A crucial portion of what is known as the standard model of particle theory, his hypotheses are expressed in mathematical terms; Higgs formulated them during the decade after he earned his doctoral degree in 1954, in an attempt to answer two questions about the nature of the universe: why do some elementary, or fundamental, particles have mass while others do not, and how did those that have mass acquire it? The answers to those questions will shed light on what happened in the infinitesimally tiny fraction of a second following the Big Bang—the moment when, theoretically, the universe, including time as well as space and matter, came into existence, 10 billion to 20 billion years ago. Higgs has proposed that, in addition to the electromagnetic and gravitational fields (whose existence has been proved experimentally or through observation and can be expressed mathematically as well), there exists a mass-generating field that pervades the whole universe. Associated with that field, he has suggested, is a particle that generates mass; that particle was dubbed (not by Higgs) the Higgs boson. The validity of Higgs's equations and the others connected with the standard model depended upon the existence of that particle, but no one had ever detected its physical presence until July 2012, at the Large Hadron Collider (LHC) near Geneva, Switzerland; prior to the construction of the LHC, even the most powerful accelerators that had been operating in recent years were insufficiently powerful to detect it. In an interview with Richard Wilson for the *London Sunday Times* (June 15, 2008), the theoretical physicist John Ellis said, "All this is important because it tells us how the universe works. People have been trying to figure out how the universe works ever since they noticed that there is a universe out there. Without Peter's work, we wouldn't have theories that made any sense." In a video of Higgs posted on YouTube.com (2009), Higgs echoed Ellis's words; if the Higgs boson turned out to be a chimera, he said, a huge body of work in contemporary physics would have to be discarded as erroneous. "Detecting the Higgs boson would represent an enormous breakthrough in particle physics," Peter Rodgers, the editor of *Physics World*, declared in that publication (July 10, 2004). Proof of the boson's existence required the painstaking analysis of an enormous amount of data produced by the LHC—as well as the repair of that machine: The collider, a project overseen by CERN, the European Organization for Nuclear Research, functioned for only two weeks before a series of technical problems forced it to be shut down. Operations were delayed until November 2009. On July 4, 2012, CERN scientists announced that the LHC's ATLAS and CMS detectors had found a particle with properties consistent with the proposed Higgs boson, an important confirmation of Higgs's theories.

Higgs taught at the University of Edinburgh in Scotland for thirty-six years, until his retirement in 1996. In 2004, when Rodgers talked with him, he seemed "embarrassed by the fame that his eponymous boson has brought," as Rodgers wrote. "In conversation he talks about 'the so-called Higgs field' and the 'so-called Higgs model,' and is keen to give credit to a host of other physicists whose work has led to our current understanding of the generation of mass within the Standard Model"—physicists including Abdus Salam, Sheldon Glashow, Steven Weinberg, Philip Anderson, Jeffrey Goldstone, Yoichiro Nambu, Francois Englert, Robert Brout, Gerald Guralnik, Carl Hagen, and Tom Kibble. (Englert, Brout, Guralnik, Hagen, and Kibble are widely considered to be codiscoverers of the boson on which the Dutch theoretical physicist Gerardus 't Hooft conferred Higgs's name in 1971.) "I get very uneasy when people try to attach too much importance to me," he said to James Morgan for the *Glasgow, Scotland,*

*Herald* (April 7, 2008). His modesty notwithstanding, Higgs has won or shared a bevy of honors, among them the Rutherford Medal (1984) and the Paul Dirac Medal (1997), both from the Institute of Physics, the main professional body for physicists in Great Britain and Ireland, for "outstanding contributions to theoretical (including mathematical and computational) physics." He has also received the Wolf Foundation Prize in Physics (2004), considered among physicists second in prestige only to the Nobel Prize.

## Early Life and Education

Peter Ware Higgs was born to Thomas Ware Higgs and Gertrude Maude Higgs on May 29, 1929, in Newcastle upon Tyne (commonly referred to simply as Newcastle), a city in northeastern England. His father's job, as a sound engineer for the British Broadcasting Corporation (BBC), forced the family to move repeatedly. For that reason Higgs missed some of his early schooling. As a youngster he developed asthma; after he suffered a bout of pneumonia, his mother home-schooled him for a while. "She was very motivated to push me," Higgs told Ian Sample for the *London Guardian* (November 17, 2007). "My father, I think he was just rather scared of children." When the BBC sent his father to work in Bedford, Higgs and his mother moved to Bristol, more than one hundred miles distant, in the southwestern part of the country.

In Bristol, Higgs attended Cotham Grammar School, where science was emphasized, and he excelled at math. The German air force's heavy bombings of Bristol during World War II sometimes interrupted his schooling, and he once broke his arm when he fell into a bomb crater near the school. Cotham was the alma mater of Paul Dirac (1902–1984), a Nobel Prize–winning theoretical physicist who was a pioneer in the fields of quantum mechanics and quantum electrodynamics. Complex mathematical calculations, not physical experiments or observations, led Dirac to predict—correctly, as later discoveries proved—that for every subatomic particle, there exists a mirror twin, an antiparticle, with the same mass but opposite charge. Dirac's work inspired Higgs to study theoretical physics. "It's about understanding!" Higgs exclaimed to Sample. "Understanding the world!"

When he was seventeen, Higgs enrolled at the City of London School, an independent day school known both for its diverse student body and its strong science and mathematics curricula. During his high-school years (and later at college), he much preferred, and was far more skilled at, grappling with mathematical formulas than handling equipment and chemicals or other materials in science labs. Having accepted his parents' contention that Great Britain's most prestigious universities—Oxford and Cambridge—"were all very well for the children of the idle rich to go and waste their time and that of their tutors" and that "if you were serious about university, you went somewhere else," as he recalled to Sample, Higgs opted to attend King's College London, a division of the University of London. He earned a BS degree with first-class honors in physics in 1950 and an MS degree, also in physics, the following year. He worked toward his doctorate under the supervision of two theoretical chemists—Charles Coulson and H. Christopher Longuet-Higgins, a former student of Coulson's. Higgs was awarded a PhD in 1954. His thesis was entitled "Some Problems in the Theory of Molecular Vibrations"—"work which signaled the start of his life-long interest in the application of the ideas of symmetry to physical systems," according to a 2008 profile of him on the website of the School of Physics and Astronomy of the University of Edinburgh. In particle physics, "symmetry" has special meanings. One of them is "invariance," the property of remaining unchanged when certain conditions change, among them time, orientation in space, charge (positive or negative), and parity.

## Life's Work

For two years after he completed his graduate studies, Higgs worked at the University of Edinburgh. He had grown fond of Edinburgh during his youth, when he had hitchhiked to the Edinburgh Festival (the umbrella name for several cultural festivals that occur concurrently every summer). He spent the next year at University College London (part of the University of London) and the following year at Imperial College, also in London. From 1958 to 1960 he taught mathematics at University College London. He then returned to the University of Edinburgh, where he held the title of lecturer at the Tait Institute for Mathematical Physics for the next ten years. He was promoted to reader (an academic position in Great Britain) in 1970 and named to a "personal chair" in theoretical physics in 1980.

Meanwhile, in addition to his classroom duties, Higgs had become fascinated with quantum field theory, which describes the characteristics and behavior of atoms and subatomic particles and the forces that act on them, and he began to contemplate the origins of mass. A fundamental concept of physics, mass is the amount of matter that makes up anything that has substance, everything from atoms to parts of atoms and congregations of atoms: gases, liquids, and solids, the last including everything from grains of sand, animals and plants, and manmade objects to meteors, planets, moons, and stars. In physics, mass is not the same as weight, which is a measurement of Earth's gravitational pull on anything that has mass; according to the physicist Albert Einstein's famous equation $E=mc^2$—where "E" represents energy, "m" is mass, and "c" is the speed of light (186,000 miles or just under 300,000 kilometers per second)—energy also has mass. In the first scientific description of mass, the British mathematician and physicist Isaac Newton (1642–1727) wrote about it in terms of the force needed to move and accelerate a stationary object. According to Ian Sample, for several centuries "scientists were happy to think of mass as something that simply existed," but in the twentieth century, as more was learned about subatomic particles as well as the probable origins of the universe, they began to wonder where mass had come from.

Carried out mathematically rather than experimentally or through observation of astronomical phenomena, Higgs's work was based on research that other physicists had been conducting, some of it beginning in the 1950s. In the course of performing his calculations, Higgs manipulated equations of the physicists Abdus Salam and Sheldon Glashow. When he "added in his own field equations (and thus the Higgs bosons) to equations somewhat analogous to those of Salam or Glashow," Charles C. Mann wrote for *Smithsonian* (March 1, 1989), "he found that certain particles in the original equation behaved in an astonishing fashion. They began with zero mass and then mathematically 'ate up' other, unwanted particles in the field (ones that had made a mess of the mathematics in the theory he was investigating), emerging with mass." In interpreting that outcome, Higgs came up with the idea that at the moment the universe came into being, particles had no mass. Then, a minuscule fraction of a second later, as a result of what became known as the Higgs mechanism, an all-encompassing force field was formed and some particles acquired mass. In Sample's words, "Higgs's theory showed that mass was produced by a new type of field that clings to particles wherever they are, dragging on them and making them heavy. Some particles find the field more sticky than others. Particles of light are oblivious to it. Others have to wade through it like an elephant in tar. So, in theory, particles can weigh nothing, but as soon as they are in the field, they get heavy"—that is, they acquire mass and weight.

Discoveries in physics are often derived from analyses of broader phenomena. That was true of what Higgs has called his "one important idea," the crux of which came to him during a single weekend, as James Morgan reported, when he was thinking about the work of the physicist Jeffrey Goldstone. The first mention of his idea appeared in a communication published in *Physics Letters* (Volume 12, September 1964), a CERN journal, in which Higgs questioned conclusions drawn by Goldstone regarding "the nature of relativistic field theories with spontaneously broken symmetries," according to a brief biography of Goldstone on the website of the Massachusetts Institute of Technology, where he is now professor emeritus. Higgs's letter attracted scant attention. Higgs next sent another letter to the same CERN publication, elaborating on his findings and noting the possibility of the existence of a never-before-described, massive boson, but it was rejected "on the grounds that it did not warrant rapid publication," according to an article entitled "A Brief History of the Higgs Mechanism" posted in 2008 on the website of the School of Physics and Astronomy at the University of Edinburgh.

Higgs then sent his second letter to *Physical Review Letters*, a publication of the American Physical Society, where its worthiness for publication was reviewed by Yoichiro Nambu, whose discovery of the mechanism of what is called spontaneous broken symmetry earned him the Nobel Prize in physics in 2008. Higgs's missive had reached

*Physical Review Letters* on August 31, 1964, the same day that the journal had published, in its thirteenth volume, a paper on work (about which Higgs had been unaware) conducted by the Belgian physicists François Englert and Robert Brout, in which Englert and Brout, through very different calculations, had presented a hypothesis similar to his. At Nambu's request, Higgs added to his letter material about

---

**"People have been trying to figure out how the universe works ever since they noticed that there is a universe out there. Without Peter's work, we wouldn't have theories that made any sense."**

---

Englert and Brout's findings, after having discussed their work with two American particle physicists—Gerald Guralnik of Brown University and Carl R. Hagen of the University of Rochester—and Tom Kibble of Imperial College London. Higgs's revised second letter, entitled "Broken Symmetries and the Masses of Gauge Bosons," as well as two letters by Guralnik, Hagen, and Kibble, appeared in a subsequent issue of *Physical Review Letters* in 1964. "I don't know why I separated the work into two pieces," Higgs told Peter Rodgers for the *London Independent* (September 1, 2004). "Maybe if I had written it as one continuous paper, they would have rejected the lot."

Higgs's second paper sparked some discussion among physicists. On a lecture tour of the United States in the mid-1960s, Higgs talked about his hypotheses at universities including Harvard and at the Institute for Advanced Study, at Princeton. "I was facing audiences who at first thought I was a crackpot . . . ," he recalled to Rodgers. "They didn't realize that you could do something useful with the work." That realization came when Steven Weinberg and Abdus Salam used Higgs's equations to identify new particles, called W and Z, and demonstrated that those particles could combine two of the universe's four forces—the electromagnetic force and the weak force—into what was named the electroweak force. After Weinberg, Salam, and Glashow received the 1979 Nobel Prize in physics for their discoveries, "the

Higgs boson became part of the furniture in particle physics," as Rodgers wrote.

## The Nature of Particle Physics

If particle physics is thought of as a room, as in Rodgers's analogy, it is a room that in the twentieth century grew increasingly crowded with furniture but still lacked some important pieces and failed to present a unified whole—a situation that was still true in early 2009. Scientists now know that every atom—the smallest particle whose composition can be changed by chemical means—has a nucleus composed of one or more protons orbited by one or more electrons, and that, except for one form (or isotope) of the hydrogen atom, every atom contains one or more neutrons. (Protons have a positive charge; electrons have a negative charge; and neutrons are either totally or virtually charge-free.) The nature of every chemical element on Earth—oxygen, nitrogen, carbon, gold, mercury, iodine, and uranium are among the ninety-two naturally occurring elements—is determined by the number of protons in their atoms. Electrons are elementary, or fundamental, particles—that is, they have no component parts and therefore cannot be subdivided; protons and neutrons are composed of other fundamental particles—quarks, which are 100 million times smaller than the average atom (whose diameter, in turn, is smaller than a billionth of a meter). Scientists have identified six kinds, or flavors, of quarks: up, down, charmed, strange, top, and bottom. The top quark, which is the largest, is 4,400 times bigger than the smallest, the up quark; three of the quark types have an electric charge of +2/3 each, while the electric charge of each of the other three is −1/3. Every proton consists of one down quark and two up quarks; every neutron consists of one up quark and two down quarks. The other fundamental particle is the lepton. There are six types of leptons—the electron, the muon, and the tau particle, each of which has an associated neutrino. Furthermore, all fundamental particles can be categorized as either fermions or bosons, depending on their spin, a highly complex, intrinsic property associated with the angular momentum of the particle, and on their "quantum state," which in the theory of quantum mechanics is expressed in statistical, rather than absolute, terms. Bosons can coexist with one another in the same state at the same time and location; fermions (which include quarks and leptons) cannot coexist in the same state at the same time and location. Bosons may be fundamental particles or composite particles; the fundamental bosons are called gauge bosons (which include the photon, the fundamental unit of light and every other form of electromagnetic radiation; the W and Z particles; gluons; and, hypothetically, the Higgs boson). Scientists have further separated the approximately two hundred known fundamental or composite particles into three families. Only those in the first family exist for more than a fraction of a second; they make up the atoms of matter that we are familiar with in our everyday lives. Evidence for those in the second and third families has come from experiments conducted in atom smashers. All particles composed of quarks are called hadrons—hence the name of the LHC. There are two types of hadrons: those containing three quarks, which are called baryons, and those containing one quark and one antiquark (one of the antiparticles predicted to exist by Paul Dirac), called mesons.

According to the standard model of particle theory, there are also four fundamental forces in nature: the electromagnetic force, which acts between atomic particles carrying an electrical charge; the strong nuclear force, which acts between quarks and holds atomic nuclei together; the weak nuclear force, which operates between leptons and is associated with particle emission and nuclear decay (the processes that provide the basis for carbon dating, by which geological ages or the ages of fossils or other ancient materials are estimated); and gravitation, which operates on cosmic levels. Each of the forces is "carried" to subatomic particles, atoms, or congregations of atoms by a particular "messenger" particle, also called a force carrier or mediator. The carrier of the electromagnetic force is the photon, which has no mass and no charge; the carrier of the strong nuclear force is the gluon, of which there are eight kinds, each of which has a large mass and what is called a "color" charge; the weak force has three carriers—the positive and negative W particles and the neutral Z particle, which have large mass; that of gravitation is the graviton, which has no charge and no mass (and exists only hypothetically; no physical evidence for it has been found). Carriers with no mass move at the speed of light and act over long distances; those with large mass move only over short distances and at less than the speed of light.

## The Higgs Mechanism

The existence of the Higgs mechanism, Higgs field, and Higgs boson provides the only plausible explanation in the standard model of particle theory for the existence of mass. According to John Ellis, in an article entitled "Into a New World of Physics and Symmetry," posted online in the magazine *Symmetry: Dimensions of Particle Physics* (August 2006), most theoretical and particle physicists (one notable exception being Stephen Hawking) felt confident that the LHC would provide proof of their existence. Under construction from 1994 to 2008 at a cost of $8 billion, the LHC is located 300 to 570 feet below ground. The creation of some 10,000 scientists from many countries, it is far more powerful than what until then had been the world's most powerful atom smasher—the Tevatron collider at the Fermi National Accelerator Laboratory (Fermilab) near Chicago, Illinois. In a ring seventeen miles long surrounded by thousands of supercooled magnets (their temperature is near absolute zero: $-459.67$ degrees Fahrenheit, $-273.15$ degrees Celsius) in the LHC, beams of protons moving in opposite directions at 99.999999 percent of the speed of light will collide, reproducing conditions similar to those immediately after the big bang. The LHC is "the world's most powerful microscope," John Ellis wrote, "with a resolution thousands of times smaller than the diameter of a proton. The high energies and small distances accessible with the LHC will be similar to the conditions of the very early universe, shortly after the big bang, turning the collider into a telescope and a time machine that will reveal the physics that underlies the world around us." Those revelations may include evidence of new types of symmetries or of supersymmetry, which implies, as Brian Greene wrote for the *New York Times* (September 11, 2008), that "for every known species of particle (electrons, quarks, neutrinos, etc.)" there exists not only an antiparticle but "a partner species," or "sparticle"—selectrons, squarks, sneutrinos—or a fermion associated with every boson. The researchers at the LHC are interested in producing never-before-seen particles and answering questions about the nature of dark matter, why there are so many different kinds of particles, why the number of particles in the universe is many magnitudes greater than the number of antiparticles, why the gravitational force is so much weaker than the other fundamental forces, and whether the universe contains dimensions other than the three we know.

In 2010, Higgs received the J. J. Sakurai Prize for Theoretical Particle Physics recognition of outstanding contributions to the field of particle physics. Higgs's work is the subject of several books, among them *Reflections on the Higgs System* (1997), by Martinus Veltman; *Search for the Higgs Boson* (2006), edited by John V. Lee; *Perspectives on Higgs Physics* (2008), edited by Gordon L. Kane; and *The Quantum Frontier: The Large Hadron Collider* (2009), by Don Lincoln. The title of another book, *The God Particle: If the Universe Is the Answer, What Is the Question?* (1993), by the Nobel Prize–winning experimental physicist Leon Lederman, led some people to refer to the Higgs boson as the God particle. "I wish [Lederman] hadn't done that," Higgs has said, as quoted by Richard Wilson. "I have to explain to people it was a joke. I'm an atheist, but playing around with names like that could be unnecessarily offensive to people who are religious."

## Personal Life

Although Higgs was not directly involved in the operation of the LHC or the analyses of the data that it produces, he was present when CERN researchers announced the discovery of the Higgs boson. More than a decade before he retired, he had begun to decrease the amount of energy he devoted to research and focus more on teaching. He has attributed that change to problems in his personal life. "After the breakup of my marriage, I think I just lost touch with the things I should have been learning about just to follow up my own work. I couldn't keep up," he told a reporter for *Scotland on Sunday* (September 14, 2008). Higgs and his wife, an American linguist named Jo Ann (nicknamed Jodie), divorced in the early 1970s; the two had met in the 1950s when both were members of the Campaign for Nuclear Disarmament, a network of British organizations opposed to the spread of nuclear weapons. A fellow of the Royal Society and the Royal Society of Edinburgh, Higgs lives in Edinburgh. A shy man who treasures his privacy, he enjoys walking, listening to music, and swimming. He has two children—Chris, a computer scientist, and Jonny, a jazz musician—and is a grandfather.

## Further Reading

Morgan, James. "Back to the Future." (Glasgow, Scotland) *Herald* 7 Apr. 2008: 13. Print.

"Quiet Man Making a Big Bang." *Scotland on Sunday* 14 Sept. 2008: 19. Print.

Sample, Ian. "The God of Small Things." *The Guardian* 1 Nov. 2007: 44. Print.

Sample, Ian. "The Man Behind the 'God Particle.'" *NewScientist* 13 Sept. 2008: 44. Print.

# Iijima, Sumio

## Japanese physicist

**Born**: May 2, 1939; Koshigaya, Japan

On June 23, 1991, while examining minuscule crumbs of carbon with an electron microscope, the Japanese physicist Sumio Iijima discovered what, soon afterward, he dubbed carbon nanotubes. Iijima's description of his discovery, published in the journal *Nature* (November 7, 1991), gave a tremendous boost to the fledgling science of nanotechnology. Immediately, molecular manufacturing—the fabrication of microscopically small devices for use in medicine, communications, electronics, energy production, chemical processes, and other industries—moved significantly closer to reality, as did the possibility of "extending precise control of molecular structures to larger and larger scales," as K. Eric Drexler predicted in his book *Engines of Creation: The Coming Era of Nanotechnology* (1986). "The earthshaking insight of molecular nanotechnology is that, when we reach this scale, we can reverse direction and begin building up, making products by placing individual atoms and molecules exactly where we want them," Mike Treder wrote for the *Futurist* (January/February 2004). In addition, according to a writer for *Nature*'s online "Physics Portal" (2001), "as a tool to test quantum mechanics and model biological systems," among other areas of basic science, "nanotubes seem to have unlimited potential." For his discovery of carbon nanotubes and other contributions to science, Iijima has earned some forty awards and honors from schools, organizations, and governments in Japan, China, Europe, and the United States. He received his first prize—the Bertram Eugene Warren Diffraction Physics Award from the American Crystallography Society—in 1976, when he was a postgraduate research associate at Arizona State University. Among those he has received in recent years are the John M. Cowley Medal from the International Federation of Societies for Microscopy in 2006, the Gregori Aminoff Prize in crystallography from the Royal Swedish Academy of Sciences in 2007, the award for technical scientific research from the Prince of Asturias Foundation in Spain in 2008, the "No Boundaries" Innovation Award from the British weekly the *Economist*, in 2008, and admission to the Norwegian Academy of Science and Letters as a foreign member in 2009. Iijima was elected a foreign associate of the US National Academy of Sciences in 2007.

Named in 1974, nanotechnology involves the study, formation, and manipulation of molecules or of bits of matter only a few molecules in width. The primary unit of measurement is the nanometer: one billionth of a meter (a meter being close to 3.281 feet, or a little over a yard). The nanometer is about ten times the radius of a hydrogen atom (the smallest atom), or about the size of one carbon atom, or approximately 1/100,000th the width of a human hair. Carbon nanotubes are seamless tubes of carbon atoms joined in a honeycomb pattern; the tubes are at most a few nanometers in diameter and a few micrometers in length. (A micrometer is one-millionth of a meter.) While scientists have found tubular structures among a few minerals in their natural state, carbon nanotubes are manmade; none have ever been detected in the natural world. The carbon nanotubes that Iijima saw on that summer day in 1991 were not created intentionally; they came into being during experiments in which carbon rods were vaporized. The products of such vaporization—spheres whose surfaces are patterned like honeycombs—had drawn the attention of Iijima's colleagues in Japan, the United States, and Great Britain. The spheres were "not interesting for a microscopist," however, Iijima told Dennis Normile for *Science* (November 18, 1994). Rather, what captivated Iijima were the whiskery structures that had appeared on the ends of the electrodes used in the vaporization process; they reminded him of the "whisker-like" structure of silver-bromide crystals, which he had studied a quarter-century earlier, while conducting research as a graduate student, as he told Irene M. Kunii for *Business Week* (July 8, 2002), and the tubular structure of chrysotile asbestos, which he had examined when another researcher in his lab studied it in the 1960s. "When I saw carbon nanotubes under the electron microscope, my old experiences came immediately to mind and helped me to figure out what they were," he wrote for a Meijo University website. His path as a researcher has also been influenced by his uncommon understanding of the nanoworld, which stems from skills in electron microscopy that are second to none.

Iijima's journey to success in science began in early childhood, when, "stated very simply, I loved nature," as he wrote in "About Myself," an online essay for the NEC Corporation in Tsukuba, Japan, where he has worked since 1987 as a senior principal researcher. "I learned many things from my experiences with nature," he wrote, "and I believe that this helped me to develop both sensitivity and insight." He also wrote, "The most important thing that I learned while in the US was 'Don't do what others have done.'" Iijima "has a sense of what to look for, of what will give us the most interesting and most valuable information," Hiroyoshi Rangu (also known as Roy Lang), then the general manager of NEC's Fundamental Research Laboratories, told Normile. From 1998 to 2002, along with Christian Colliex of the French National Center for Scientific Research, Iijima codirected research for the Nanotubulites Project, a joint venture of the International Cooperative Research Project and the Japan Science and Technology Agency. In Japan, in addition to his work with NEC, Iijima has been a professor in the Department of Materials Science and Engineering of Meijo University in Nagoya since 1998; the director of the Nanotube Research Center (also known as the Research Center for Advanced Carbon Materials) of the National Institute of Advanced Industrial Science and Technology in Tsukuba since 2001; project leader of the Carbon Nanotube Capacitor Development Project of NEDO (New Energy and Industrial Technology Development Organization) since 2006; and a distinguished invited university professor at Nagoya University since 2007. He has also held the post of dean of the Advanced Institute of Nanotechnology at Sungkyunkwan University in Suwon, Gyeonggi-do, South Korea, since 2005. More than 482 papers published in scientific journals carry Iijima's name as author or coauthor.

## Early Life and Education

Sumio Iijima was born on May 2, 1939, to Fukumatsu and Take Iijima in Koshigaya, a town in the Saitama Prefecture of Japan. At the small primary school he attended—there were only three classes, with a total of 120 pupils—he had a fine science teacher "who influenced me very much," he told Irene M. Kunii. His primary interests were unrelated to book learning, though; as a boy, he recalled in his NEC essay, "I collected plants and insects, I fished, and I kept a menagerie of small animals, including pigeons, rabbits, snakes, frogs, and crabs." As a high-school student, he ranked in the "lower middle" of his class, because he did not study diligently. "You see, I had so many things to do," he explained to Kunii. "I loved sports, collecting minerals, and looking at the stars at night." He was also a member of his school's mountaineering and music clubs. Iijima failed his university entrance exams on his first try. He then spent a year learning to play the mandolin and preparing to retake the exams, which, much to his dislike, required much memorization. His retest mark was sufficient to gain him acceptance into the University of Electro-Communications in Tokyo. He concentrated in communications engineering until his final year, when he changed his major to chemistry. When he realized that he wanted to gain admission to a graduate school, he uncharacteristically applied himself to his studies wholeheartedly. He earned a BS degree in 1963 and then enrolled at the Graduate School of Physics at Tohoku University in Sendai, Japan. "It just so happened that I was assigned to the laboratory of Professor Tadatoshi Hibi," a pioneer in the use of electron microscopy for research, he wrote in "About Myself." (Hibi had built Japan's first electron microscope in 1936.) That assignment "determined the rest of my life," Iijima told Kunii. He recalled in his essay, "It wasn't that I had a particularly strong desire to do research with electron microscopes at the time, but I found that I was perfectly suited to research in this field." He also wrote that the change in the trajectory of his career, from physics to microscopy, and "the fact that I dove into new fields," were "a result of my determination to find something"—that is, something new to science. Iijima earned an MS degree in chemistry in 1965 and a PhD in solid-state physics in 1968 from Tohoku University.

### Life's Work

From 1968 to 1975 Iijima worked as a research associate at Tohoku's Research Institute for Scientific Measurements. Also during that time, for twelve years beginning in 1970, he engaged in research at the Center for Solid State Science at Arizona State University, in Tempe, under the physicist John M. Cowley, another major figure in electron microscopy and crystallography. "It was great because I had complete freedom and a very good microscope," he told Kunii. Iijima constructed a more sophisticated electron microscope and became the first microscopist ever to capture an image of individual atoms (those

> **"When you do not have any clue as to how to start new research, you cannot rely on anyone but yourself. What you can rely on when you face a serious difficulty is nothing but your experience."**

of tungsten). Colleagues told him that he "had achieved the dream held by researchers since the invention of the electron microscope in 1932—that of actually seeing a single atom," he recalled in "About Myself." With Cowley, Iijima wrote a paper titled "Electron Microscopy of Atoms in Crystals," published in *Physics Today* (March 1977); according to its deck, "Now we can see atoms in crystals directly with electron microscopy, allowing us to determine structures of both ordered and disordered solids and to study the way atoms cluster around crystal defects." (Ordered solids, or crystals, have precise, regular patterns of atoms or molecules, called lattices; their lattices may have defects, but only very few, and far between. Disordered solids have random patterns or crystal lattices with many defects.) With Michael A. O'Keefe and Peter R. Buseck, members of the University of Arizona Departments of Chemistry and Geology, Iijima set out to expand the boundaries of what could be seen with electron microscopes by improving their resolution. Their influential paper on their work appeared in *Nature* (July 27, 1978). The abstract read, "It is now possible to compute images of crystal structures corresponding to experimental images produced by using high-resolution transmission electron microscopy (HRTEM)," and it noted that the authors had used several minerals and a binary oxide in their investigations.

In 1979, as a visiting senior scientist, Iijima joined the Department of Metallurgy and Materials Science at Cambridge University in England. While studying graphite, a form of carbon, Iijima noticed a previously undescribed spherical arrangement of carbon atoms. His article about it contained only a partial description of its structure. Six years later the British scientist Harold Kroto and the American scientists Robert Curl and Richard Smalley published in *Nature* (November 1985) a description of $C_{60}$ (C is the symbol for carbon), in which carbon atoms are joined to form a ball. The men christened it buckminsterfullerene because it reminded them of the geodesic dome designed by Buckminster Fuller. Iijima recognized buckminsterfullerene as the spherical structure he had seen years earlier. His second paper about the structure, "The 60-Carbon Cluster Has Been Revealed!," written at Smalley's request and published in the *Journal of Physical Chemistry* (Volume 91, 1987), provided corroborating evidence that a new form, or allotrope, of carbon had been discovered. (Kroto, Curl, and Smalley won the 1996 Nobel Prize in chemistry for their work.) Later, additional allotropes of carbon were found; they became known collectively as fullerenes (or, less commonly, buckyballs). Despite the efforts of many researchers, the anticipated practical applications of fullerenes have not yet materialized.

In the meantime, in 1982 Iijima had returned to Japan to become a group leader in the Hayashi Ultra-Fine Particle Project, an activity of the national Exploratory Research for Advanced Technology (ERATO) program. "My major task in this project was to develop a new high resolution electron microscope accessible for characterizing ultra-fine crystals without exposure to air," he wrote for nanocarb. meijo-u.ac.jp. With that tool he demonstrated, as he reported in the *Journal of Electron Microscopy* (December 1985), that ultrafine, crystalline particles of gold "move about like amoeba," in his words; both their external form and internal arrangements of atoms change continuously. He earned the 1985 Nishina Memorial Award for that finding.

In 1987 Iijima left the ERATO project and signed on as a senior principal researcher with the NEC Corporation. He chose to enter the private sector for the first time, at the age of forty-eight, partly because private facilities like NEC's "have unique materials produced using expensive equipment that cannot be made in universities, and I was sure that these materials would become the basis for exceptional research results using electron microscopes," he wrote in "About Me." In addition, NEC executives promised to give him the resources to build an ultrahigh-vacuum, high-resolution transmission electron microscope. He also liked NEC's style of management, which, he wrote, "was based on the approach of 'broad proliferation of science and technologies, and the creation of new value,' and of 'giving back to society.'"

## Discovery of Carbon Nanotubes

Iijima was using not the ultrahigh-vacuum electron microscope but a standard electron microscope on June 23, 1991, when he obtained the clearest pictures yet of the structures that would come to be known as carbon nanotubes. His achievement was a combination of instinct, skill, and luck. "I emphasize that it was serendipity," Iijima told Dennis Normile, adding, "I have the best technique in microscopy." His paper about his findings, titled "Helical Microtubules of Graphitic Carbon," attracted worldwide attention among chemists, physicists, and others when it appeared in *Nature* (November 7, 1991). The first nanotubes that he detected had several concentric walls; in 1993 he discovered single-walled carbon nanotubes. In either case, because of their smallness, nanotubes "exhibit unique physical and chemical properties," as he wrote for nanocarb.meijo-u.ac.jp. "A single piece of single-wall carbon nanotube can be a transistor . . . ," he continued. "Many other industrial applications utilizing the unique properties of carbon nanotubes include the electron emitter source with high current density, highly conductive electrical wire, the high thermal conductor for the heat radiator, the probe needle for scanning probe microscopes, molecular sieves, gas absorbers, [and] carriers for drug delivery systems in nano-bio medicine." Nanotubes can be made to function either as metals or semiconductors and are the smallest known molecules to display metallic properties. The tubes' conductivity of electricity is higher than that of copper, and their conductivity of heat is higher than that of diamond (whose heat-conductivity ability is superior to that of tungsten, beryllium, aluminum, gold, copper, and silver). They are also extremely strong (reportedly, fifty times stronger than an equivalent piece of steel) and may be used to increase the strength of various materials. For example, some manufacturers of tennis rackets now incorporate carbon nanotubes in the yoke of the frame to strengthen it.

Iijima has published hundreds of papers describing his research on nanotubes. They cover subjects including nanotubes' growth and structural flexibility; their formation by such means as laser evaporation and hydrogen-arc discharge; responses of nanotube bundles to visible light; effects on them of oxygen, heat, and high pressure; techniques for opening and filling them; methods of mass-producing and purifying them; and methods of fabricating carbon-nanotube tips for use in scanning-probe microscopy or atomic-force microscopy. The titles of a handful of papers hint at the scope of Iijima's work: "A Simple Way to Chemically React Single-Wall Carbon Nanotubes with Organic Materials Using Ultrasonication" (*Nano Letters*, 2001); "Metal-Free Production of High-Quality Multi-Wall Carbon Nanotubes, in Which the Innermost Nanotubes Have a Diameter of 0.4nm [nanometers]" (*Chemical Physics Letters*, 2002); "Diameter Enlargement of Single-Wall Carbon Nanotubes by Oxidation" (*Journal of Physical Chemistry*, 2004); "Water-Assisted Highly Efficient Synthesis of Impurity-Free Single-Walled Carbon Nanotubes" (*Science*, 2004); "Synthesis of Single- and Double-Walled Carbon Nanotube Forests on Conducting Metal Foils" and "Metallic Wires of Lanthanum Atoms Inside Carbon Nanotubes" (*Journal of the American Chemical Society*, 2006 and 2008, respectively).

Iijima has also studied nanobeads and nanohorns, the latter of which resemble chubby, single-walled carbon nanotubes closed on one end with what looks like a cone-shaped cap (the horn); various sources describe them as resembling dahlias in shape. In an interview in Japanese on July 8, 2003, for NanoNet.go.jp (posted in English on that site on July 22, 2004), Kuniko Ishiguro wrote that Iijima "now thinks that commercial applications of carbon nanohorns will be realized much earlier than those of carbon nanotubes." Iijima told Ishiguro, "People are exaggerating carbon nanotubes too much. However, I can say with confidence that carbon nanotubes have made great contributions to basic science." He also said, "I had conducted research using electron microscopy for thirty years before I discovered carbon nanotubes, so discovering them is just one of the results of my research based on electron microscopy." And he offered some advice to aspiring scientists: "When you do not have any clue as to how to start new research, you cannot rely on anyone but yourself. What you can rely on when you face a serious difficulty is nothing but your experience."

Iijima's name entered the *Guinness Book of World Records* in connection with a lecture he gave in May 1997 at the Friday Evening Discourse at the Royal Institution in London, England. Lectures at that event, which dates from 1825 and is open to the public, start promptly with the sound of a bell and must end—even if the speaker is in midsentence—when the bell rings again, precisely one hour later. Iijima ended his talk three seconds before the second bell rang.

## Personal Life

Iijima's recreational interests include playing the flute, playing tennis, and skiing. He and his wife, Aida Nobuko, a professor of gerontological nursing at Nagoya University, married in 1968. They live in Nagoya and have two children, Masako and Arihiro.

## Further Reading

"Benjamin Franklin Medal Awarded to Dr. Sumio Iijima, Director of the Research Center for Advanced Carbon Materials, AIST." aist.go.jp. AIST, Feb. 2002. Web. 7 Aug. 2012.

Normile, Dennis. "A Sense of What to Look For." *Science* 18 Nov. 1994: 1182. Print.

"Sumio Iijima." *BusinessWeek*. Bloomberg L.P., 7 July 2002. Web. 7 Aug. 2012.

"Sumio Iijima." *Nano Tsunami*. Voyle.com. Web. 7 Aug. 2012.

# Jin, Deborah

## American physicist

**Born**: November 15, 1968; Palo Alto, California

In 1999 the world-renowned journal *Science* named as one of the top ten scientific breakthroughs of the year an ingenious experiment conceived and executed, with some assistance, by the physicist Deborah Jin. Overcoming hugely difficult obstacles connected with the nature of fundamental particles, Jin and her graduate student Brian DeMarco succeeded in cooling atoms of an isotope of the element potassium to a temperature significantly closer to absolute zero than had ever been done before with any material of its kind. Absolute zero is defined as –459.67 degrees Fahrenheit, or –273.15 degrees Celsius (or centigrade), or zero degrees on the Kelvin scale. At absolute zero, all the elementary particles that constitute matter are at what David H. Freedman in *Discover* described as "the stillest possible state," with no motion "except for a minimum residual buzz" (February 1993). By lowering the temperature of their experimental material to less than one-millionth of a degree above absolute zero on the Kelvin scale, Jin and DeMarco made possible a better understanding of fundamental particles and paved the way for the creation of ever more accurate and useful atomic clocks and other devices. Jin's work earned her several prestigious honors, among them, in 2000, the Presidential Early Career Award for Scientists and Engineers, the highest honor given by the US government to young scientists; in 2002, the Maria Goeppert-Meyer Prize from the American Physical Society and the National Academy of Sciences Award for Initiatives in Research; and in 2003, a $500,000 MacArthur Fellowship.

Jin, who has worked at the Joint Institute for Laboratory Astrophysics (JILA) in Boulder, Colorado, since 1995, has continued to refine and improve her techniques. In January 2004 she and two of her colleagues announced their successful cooling of their experimental material to a temperature of 50 billionths of a degree above absolute zero on the Kelvin scale and the resultant formation of a so-called fermionic condensate—"a long-sought, novel form of matter," according to a news release (January 28, 2004) from the University of Colorado at Boulder (UC-Boulder), which helps to support JILA. The news release continued, "Physicists hope that further research with such condensates eventually will help unlock the mysteries of high-temperature superconductivity, a phenomenon with the potential to improve energy efficiency dramatically across a broad range of applications." In November 2002 the popular science magazine *Discover* included Jin in its list of the fifty most important women in science.

### Early Life and Education

Deborah Shiu-lan Jin was born on November 15, 1968, in Palo Alto, California, and raised in Indian Harbor Beach, Florida. In an interview for the *Princeton Alumni Weekly* (November 19, 2003), she said that, as a scientist, she has had several mentors, but only one among them was female: her mother, who held a master's degree in engineering physics and worked as an engineer. Jin attended Princeton University in New Jersey. She spent the summer following her sophomore year as a federal-government researcher at the Goddard Space Flight Center in Greenbelt, Maryland. "That summer pretty much settled things," she told T. R. Reid for *Newsbytes* (October 7, 2003). "I think I knew from that point on that I was going to be a physicist." In her senior year Jin won Princeton's Allen G. Shenstone Prize for her exemplary academic performance in physics. She graduated magna cum laude from the school in 1990, with a bachelor's degree in physics. From 1990 to 1993 she held a National Science Foundation Graduate Fellowship in Physics. She earned a PhD from the University of Chicago in Illinois in 1995; her dissertation was entitled "Experimental Study of the Phase Diagrams of Heavy Fermion Superconductors with Multiple Transitions."

### Bose-Einstein Condensate

Also in 1995 Jin secured a position as a National Research Council research associate at the Joint Institute for Laboratory Astrophysics. That same year her supervisor at JILA, Eric Cornell, and his colleague Carl Wieman announced that they had created a new state of matter called a Bose-Einstein condensate (BEC). (In 2001 Cornell and Wieman, along with the Massachusetts Institute of Technology physicist Wolfgang Ketterle, who had formed a BEC independently, shared the Nobel Prize in Physics for their work.) BECs are named for the Indian physicist Satyendra Nath Bose and the German-born physicist Albert Einstein. According to a UC-Boulder news release (November 20, 2003), "BECs have been described as a magnifying glass for quantum physics, the basic laws that govern the behavior of all matter." In 1925, building upon Bose's work, Einstein predicted that if a dense gas were cooled to absolute zero, the atoms would clump into a sort of superatom in which the identities of the individual atoms would disappear. The superatom, Einstein predicted, would constitute a new form of matter, with properties unlike those of any known substance. Essential to the concept of BECs is the division of all fundamental particles (particles with no internal substructure) into two types: bosons (named in honor of Bose) and fermions (named for the Italian-born physicist Enrico Fermi). The characteristics that distinguish bosons from fermions are described in terms connected with the theory of quantum mechanics (the science of physics at the scale of atoms), according to which all fundamental particles have wavelike properties, and, conversely, light waves sometimes exhibit particle-like properties. Particles emit and absorb energy in discrete packets, called quanta and, in revolving about the nucleus of an atom, electrons can move from one orbit to another, with each orbit associated with a specific energy level. According to quantum theory, all particles have spin, a highly complex, intrinsic property associated with the angular momentum of the particle. The spins of bosons (which include photons, the carriers of light; gluons, the particles that bind quarks to one another; liquid helium; and W and Z bosons, the carriers of what is known as the weak nuclear force) are measured in terms of integers (for example, one, two, three); the spins of fermions (which include quarks, the building blocks of the subatomic particles neutrons and protons; leptons; and neutrinos) are measured in terms of odd numbers of half-integers (for example, $1/2$, $3/2$, $5/2$). Bosons are said to exhibit what is called Bose-Einstein statistics: they can have the same quantum state in the same place at the same time; fermions, by contrast—constrained by what is known as the Pauli Exclusion Principle—cannot coexist in the same state at the same time and location. Bosons, as the UC-Boulder news release put it, "are inherently gregarious; they'd rather adopt their neighbor's motion than go it alone," while fermions "are inherently loners" (January 28, 2004). In their natural states, to offer crude analo-

> **"This makes me optimistic that the fundamental physics we learn through fermionic condensates will eventually help others design more practical superconducting materials."**

gies, bosons may behave like two dozen cupcakes stuffed into a child's lunchbox; fermions, by contrast, resemble people standing on each step of a narrow staircase. When cooled sufficiently, Enrico Fermi predicted, fermions would create "stacks" of quantized energy states and form a vapor, called a Fermi gas; in that state, the atoms would "degenerate" and act more like waves than like particles.

In creating Bose-Einstein condensates, Eric Cornell and Carl Wieman worked with atoms of an isotope of the element rubidium (rubidium-87). First, they cooled the rubidium to 20 millionths of a degree above absolute zero by means of a laser trap (a weave of light waves generated by a half-dozen lasers); then they cooled the atoms even further, to 20 billionths of a degree above absolute zero on the Kelvin scale, by means of a magnetic trap. Thanks to their bosonic properties, when the atoms were subjected to Cornell and Wieman's supercooling techniques and reached the lower temperature, they behaved as Bose and Einstein had predicted: they all descended into the same quantum state (a near standstill), thus losing their separate identities and coalescing into a superatom—a Bose-Einstein condensate.

## Life's Work

Aiming for a similar though theoretically far more difficult result, Jin and Brian DeMarco (currently an assistant professor of physics at the University of Illinois at Urbana-Champaign) chose to work with a gas made up of potassium-40 atoms, which exhibit fermionic rather than bosonic properties. Brilliantly adapting and building upon the methods used by Cornell and Wieman, they first supercooled room-temperature potassium-40 gas by means of a magneto-optical trap, which consists of both laser beams and a magnetic field. Then, using a highly sophisticated microwave technique, they separated atoms with higher energy levels from those with lower energy levels; the former were forced to fly away, so to speak, leaving the others at a lower temperature: less than one-millionth of a degree Kelvin above absolute zero. At that point, as Jin explained on her personal page on the UC-Boulder website, "quantum mechanics starts to dominate the properties of the gas." It is not yet possible to provide visual evidence of the stacked molecules, but after the magnet was turned off and the gas expanded, it became possible to measure certain effects: what Jin termed "the thermodynamics and collisional dynamics." Those effects constituted evidence of the degenerate state of the gas. In *Science* (September 10, 1999), in a paper entitled "Onset of Fermi Degeneracy in a Trapped Atomic Gas," Jin and DeMarco described the result of their experiments. "The creation of a Fermi degenerate gas is a major scientific achievement and a lot of scientists have been trying to make it ever since we created the Bose-Einstein condensate," Carl Wieman declared, according to a UC-Boulder news release (September 9, 1999). He then predicted, "It will probably be at the top of the list of important physics news for this year." In addition to being named one of the top ten scientific achievements of the year by *Science*, Jin's work was honored by the Office of Naval Research, which in 1999 named Jin an ONR Young Investigator. In 2001 she won the National Institute of Standards and Technology's (NIST's) Samuel W. Stratton Award for "her pioneering creation of a degenerate Fermi gas in a dilute atomic vapor, a microscopic model for important scientific and technological materials."

On the website of *Nature* on November 26, 2003 (and in the print version of the journal released one week later), Jin and her colleagues Cindy A. Regal and Markus Greiner announced their successful formation of the world's first "molecular Bose-Einstein condensate from a Fermi gas," as the title of their paper described it. Working again with gaseous potassium-40, they manipulated the atoms so that the attractions between pairs of atoms increased. When this was done at a sufficiently low temperature (150-billionths of a degree Kelvin above absolute zero), loosely bound bosonic molecules formed, constituting a Bose-Einstein condensate. Five months later Jin created and observed a Bose-Einstein condensate of those molecules. The results of Jin's research were submitted for publication on the same day that a group of physicists at the University of Innsbruck at Austria announced that they had accomplished the same thing using lithium atoms; the Innsbruck group's paper appeared one month earlier than that of Jin's. "Both of these papers represent a very large step in what people have wanted to do for a long time," Eric Cornell told Kenneth Chang for the *New York Times* (November 25, 2003). Cornell described the transformation of fermion atoms into bosonic molecules as "a lovely unification of two things that in the physics world we're used to thinking of as different as male and female."

About two months later, in *Physical Review Letters* (January 24, 2004), Jin, Regal, and Greiner reported yet another achievement: instead of using magnetic fields to bind two fermions into a molecule and condense into a BEC, they used the field to create an attractive force that could not cause two fermions to form ordinary molecules but that nonetheless caused the fermions to form a condensate. This condensate more closely resembled what happens in a superconductor (in which electrons form what is known as Cooper pairs) than what happens in a BEC. Their experiment—which Eric Cornell called "a technological and scientific tour de force," according to Charles Seife in *Science* (January 28, 2004)—proved that a condensate can have properties related to both a superconductor and a BEC. "The strength of pairing in our fermionic condensate, adjusted for mass and density, would correspond to a room temperature superconductor," Jin explained, according to a UC-Boulder press release (January 28, 2004). "This makes me optimistic that the fundamental physics we learn through fermionic condensates will eventually help others design more practical superconducting materials."

Jin currently holds three titles: she is a JILA fellow; a fellow at the Quantum Physics Division of NIST (the arm of the US Department of Commerce that, together with UC-Boulder, supports JILA); and a professor adjoint with the Physics Department at UC-Boulder. Since 1996 she has lectured as an invited speaker at more than one hundred conferences or other events around the world. Jin was elected to the US National Academy of Science in 2006 and she received the Benjamin Franklin Medal in Physics in 2008. In 2011, Jin shared the Department of Commerce Gold Medal with researcher Jun Ye "for their seminal work on ultracold molecules and cold chemistry."

## Personal Life

Jin is married to the physicist John Bohn, a JILA fellow and a research scientist in the University of Colorado's Physics Department. Bohn's work is closely connected with Jin's and, along with others, they have coauthored several professional papers together. By her own account, Jin is glad that, although she works on the campus of a university, she is not required to teach; as she explained to T. R. Reid, "I'm sort of isolated from the academic politics, and being a federal employee frees you up from the teaching load and the other requirements they have for [university] faculty. I don't have to wait the six years to find out if I'm going to get tenure. The government just leaves you alone to do your work." Jin and Bohn, who live in Boulder, have one child—a daughter, Jaclyn, born in October 2002; photos of her with her parents appear on Bohn's personal page on UC-Boulder's website. "Having an infant around, I don't spend as much time at work [as I used to]" Jin told the *Princeton Alumni Weekly* interviewer. "And gosh, when you have an infant, no other job seems as difficult! There's so much at stake, and it's really hard. It gives you perspective. Like physics, it's very challenging, at least for me. I'm not a natural at it. But, like physics, it's also very rewarding."

## Further Reading

Hadhazy, Adam. "Deborah Jin Keeps It Cool With Quantum Mechanics." *Scientific American.* Nature America, 4 May 2009. Web. 9 Aug. 2012.

Parker, Ginny. "A Moment With…" *Princeton Alumni Weekly* 19 Nov. 2003. Print.

Reid, T. R. "Hot Work, Low Temperature." *Washington Post* 7 Oct. 2003: A23. Print.

"Superfluids: An Interview with Dr. Deborah Jin." *ESI Special Topics.* Research Services Group of Thomson Scientific, July 2005. Web. 9 Aug. 2012.

# Koshiba, Masatoshi

Japanese physicist

**Born**: September 19, 1926; Toyohashi, Japan

On October 8, 2002, the Royal Swedish Academy of Sciences awarded the Nobel Prize in Physics, with one half presented to Masatoshi Koshiba and Raymond Davis Jr. "for pioneering contributions in astrophysics, in particular for the detection of cosmic neutrinos," as noted on the Nobel e-Museum website. The other half of the prize was awarded to the Italian-born physicist Riccardo Giacconi of Associated Universities Inc. in Washington DC, "for pioneering contributions to astrophysics, which have led to the discovery of cosmic X-ray sources." According to the Nobel e-Museum, "This year's Nobel Laureates in Physics have used these very smallest components of the universe to increase our understanding of the very largest: the sun, stars, galaxies and supernovae. The new knowledge has changed the way we look upon the universe." Upon learning of the honor, Koshiba told Kozo Mizoguchi for the Associated Press Worldstream (October 8, 2002), "All I can say is I'm so happy, I can't find any better words." He added, "This wonderful outcome was only possible because of my young assistants' hard work. I hope the achievement can encourage young scientists and contribute to the further progress of basic physics in Japan."

## Early Life and Education

Masatoshi Koshiba was born on September 19, 1926, in Toyohashi, a city on the southeastern edge of Japan's Aichi Prefecture. His father had a career in the military, and for much of his childhood Koshiba dreamed of following in his father's footsteps. As he approached the age for military school, however, he suffered a bout of polio that prevented him from taking the entrance examinations. His interest in science was sparked while he was recovering from his illness, when a teacher gave him a book about the physicist Albert Einstein's theories.

Koshiba attended Daiichi Koto Gakko, a three-year, state-run high school located in the city of Yokosuka, west of Tokyo, where he lived in a dormitory on campus. (The school has since been incorporated into Tokyo University.)

Yoshiro Hayashi, Japan's former finance minister and one of Koshiba's classmates at Daiichi Koto Gakko, recalled him as an enthusiastic and curious student, eager to learn about many different subjects. "I remember [Koshiba] had an air you didn't see in science majors because he liked to talk about topics like literature," Hayashi told a reporter for the *Tokyo Daily Yomiuri* (October 10, 2002). "He was very bright." Despite his intelligence, Koshiba was generally a poor student. At one point, Koshiba recalled to Kwan Weng Kin for the *Singapore Straits Times* (October 10, 2002), he overheard one of his teachers declare, "Koshiba's physics results are no good. There's no way he can study physics." He took the judgment as a challenge and became more determined than ever to pursue physics at the university level.

After two attempts, Koshiba was accepted to the University of Tokyo. Yet again, his grades did not reflect his academic potential. Because his father had been dismissed from the military following World War II, Koshiba was forced to take on several part-time jobs to support his family. He subsequently missed many classes. According to his own account, he graduated from the university's science department near the bottom of his class, earning top marks in only two of the sixteen subjects he studied throughout his final two years of undergraduate school; both classes were laboratory courses, in which, according to Koshiba, "any student could earn an A as long as he or she has an adequate attendance record," as quoted by the *Daily Yomiuri* reporter. In 2002, upon winning the Nobel Prize in Physics, Koshiba released his undergraduate transcripts in order to demonstrate that anything can be possible with hard work. "There are things in the world you can achieve despite poor academic records," Koshiba said, "although I'm not saying those who have good grades should slack off. What counts most is adopting an active attitude toward studying." After graduating from the University of Tokyo, in March 1951, Koshiba applied for post-graduate study at the University of Rochester, in New York.

He drafted his own recommendation for his professor to sign, stating, "His results are not good, but he is not that stupid." Koshiba was accepted to the University of Rochester's physics program and obtained his PhD in physics, in June 1955.

## Life's Work

The next month Koshiba started work as a research associate in the physics department at the University of Chicago. He returned to Japan in 1958, when he was offered an associate professor position at

> **"There are things in the world you can achieve despite poor academic records, although I'm not saying those who have good grades should slack off. What counts most is adopting an active attitude toward studying."**

the University of Tokyo's Institute of Nuclear Study. From 1959 to 1962 Koshiba took a leave of absence from the University of Tokyo to join the University of Chicago as a senior research associate, with the honorary rank of associate professor. (He also served as acting director at the University of Chicago's Laboratory of High-Energy physics and Cosmic Radiation during this same period.) From November 1963 to February 1970 he served as an associate professor at the University of Tokyo's physics department. In March 1970 he became a full professor in the physics department, where he remained until March 1987, and served as director of a number of university research laboratories. From April to August 1987 he worked as a visiting professor at Deutsches Elektronen Synchrotron (DESY), the largest research center for particle physics in Germany, and the University of Hamburg, also in Germany. At that time he accepted another teaching position, at Tokai University in Shimizu, Shizuoka, a job he held until 1997. While at Tokai University he also served as a guest professor at CERN, the world's largest particle physics laboratory; a distinguished visiting professor at the University of Chicago; a Regent's lecturer at the University of California, Riverside; a Sherman Fairchild Distinguished Scholar at the California Institute of Technology (Caltech) and George Washington University; and a director for the Washington Liaison Office of the Japan Society for Promotion of Science. Since 1987 Koshiba has also been professor emeritus at the International Center for Elementary Particle Physics at the University of Tokyo.

At the start of his research career, Koshiba conducted numerous experiments in the field of cosmic-ray physics, which examines the physical properties of high-energy atomic nuclei entering Earth's atmosphere from outer space. By the mid-1970s he had shifted his experimental work toward proton decay, the process by which protons decay over vast amounts of time (a proton half-life in years theoretically equals 10 to the 235th power) and release energy in the form of Cherenkov light (often referred to as Cerenkov light). Physicists believed that if they could demonstrate this process of decay, they could predict how the universe itself may decay and finally end. In 1983 Koshiba finished installing a proton-decay detector in an underground mine in Kamioka from which zinc had been extracted for 1,000 years. (The experiment had to be held deep underground, so that other forms of cosmic radiation would not interfere.) His equipment consisted of a large tank filled with 2,140 metric tons of very pure water surrounded by 1,100 20-foot photomultiplier tubes to detect the Cherenkov light emitted by protons. His research team began gathering data from the Kamiokande detector—as it had been named—on July 6, 1983.

## The Solar Neutrino Problem

As Koshiba explained to Charles C. Mann and Robert P. Crease for *Science* (March 1986), the study of proton decay was complicated by interference from solar neutrinos, the chargeless subatomic particles emanating from the sun that continually bombard Earth. "The most important obstacle to proton decay experiments is the neutrino background," he said. The mysterious particle known as the neutrino had been predicted in 1930 by the Swiss physicist Wolfgang Pauli (who won the Nobel Prize in 1945), but it took another twenty-six years for scientists to prove its existence. In 1920 the British astrophysicist Sir Arthur Eddington proposed that the sun's energy was generated by nuclear reactions that transformed hydrogen into helium. (By that time it was widely accepted that the sun was composed of mainly hydrogen and helium, although since the mid-nineteenth century it had been assumed that the sun's energy came from its gravitational collapse.) The nuclear reactions, according to Eddington's theory, would produce two small particles known as neutrinos for each helium nucleus formed. The neutrinos would then radiate through space; it is now estimated that many billions of neutrinos pass through each square centimeter of Earth every second. Since they interact very weakly with matter, their presence is virtually undetectable.

In 1946 the Italian physicist Bruno Pontecorvo suggested that occasionally the sun might produce high-energy neutrinos that could interact with a chlorine atom, forming a radioactive isotope of argon. Applying this theory, the American physicist Raymond Davis Jr. of the Brookhaven National Laboratory in Upton, New York, became the first to undertake the difficult task of verifying the existence of solar neutrinos. In 1965 he conducted experiments in the Homestake Gold Mine in Lead, South Dakota. He developed a unique method for extracting the argon atoms produced by the interaction of neutrinos and chlorine and in 1968 published results confirming the existence of solar neutrinos. However, his results affirmed the presence of only one-third of the neutrinos predicted. Accounting for this discrepancy became known as the solar neutrino problem.

While Davis continued to refine his work at the Homestake tank—his experiment ran almost continuously from 1970 to 1994, when he stopped collecting data after counting approximately 2,000 argon atoms—Koshiba was making his own contribution to neutrino research. Although the Kamiokande experiment was originally designed to study proton decay, Koshiba had become inspired by Davis's results and began exploring the solar neutrino problem. When neutrinos passed through the Kamiokande tank, they interacted with atomic nuclei in the water, releasing electrons and, at the same time, small flashes of blue Cherenkov radiation. By using his ultrasensitive photomultiplier tubes, Koshiba was able to observe these flashes, thus detecting the presence of neutrinos. Because his detectors could register exactly when each neutrino passed through the tank, as well as the direction from which it came, he was able to confirm that the neutrinos were in fact coming from the sun. He further corroborated Davis's work by reporting similar deficiencies in the number of neutrinos observed.

On February 23, 1987, the Kamiokande detector recorded an enormous burst of neutrinos from a supernova explosion named 1987A in the Large Magellanic Cloud (LMC), a galaxy neighboring the Milky Way and estimated to be about 169,000 light years from Earth. When a neutron star forms following a supernova explosion, most of the star's energy is released into the universe in the form of neutrinos. (A neutron star is the dense, imploded core of an exploded star.) Koshiba's research team was able to observe twelve of the quadrillions of neutrinos passing through the detector. The experiment proved to be the first time that neutrinos were captured from outside the solar system.

In the mid-1990s Koshiba began construction on a larger neutrino detector known as Super Kamiokande, which was completed in 1996. Built on the site of the earlier Kamiokande facility, Super Kamiokande contained 50,000 metric tons of water and over 10,000 photomultipliers, making it the largest neutrino detector in the world. In June 1998 Koshiba's research team revealed two years of data that provided evidence of a new phenomenon known as neutrino oscillation, in which the neutrinos appeared to change form, or, as dubbed by scientists, flavor. One important implication of the findings was that neutrinos might not, as was previously believed, have zero mass. Secondly, the data provided a possible explanation for the lower than expected numbers of solar neutrinos detected in earlier experiments, such as Davis's: the neutrinos could possibly have changed on their way to Earth into undetectable forms. The theory was confirmed in 2001 by researchers at the Sudbury Neutrino Observatory in Ontario, Canada, who were able to detect three types of neutrinos; in addition to the electron neutrinos found in earlier experiments, the Sudbury researchers found two non-electron varieties, called muon and tau neutrinos. They also found further evidence to support Koshiba's findings that neutrinos do, in fact, have mass. This significant discovery has required a modification of the Standard Model for subatomic particles. (The Standard Model of particle physics is a theory that describes the fundamental forces, as well as the fundamental particles, that make up all matter.) In accounting for the mass of neutrinos and other subatomic particles, scientists must also reexamine their theories regarding the collective mass of the universe. This intensive new field of research is known as neutrino astronomy.

In addition to the Nobel Prize, Koshiba has received numerous prestigious honors and awards, including der Grosse Verdienstkreutz from the president of the Federal Republic of Germany (1985), the Nishina Prize from the Nishina Foundation (1987), the Asahi Prize from the Asahi Press (1988 and 1999), the Order of Culture from the Japanese government (1988), the Academy Award from the Academy of Japan (1989), the Fujiwara Prize from the Fujiwara Science Foundation (1997), and the Order of Cultural Merit conferred by the Emperor of Japan (1997). Koshiba, along with Davis, also received the 2000 Wolf Prize from the state president of Israel; the two physicists shared the award's $100,000 prize. In 2002 he was a recipient of the W. K. H. Panofsky Prize from the American Physical Society, along with Japanese physicists Takaaki Kajita and Yoji Totsuka. Koshiba currently serves as a councilor for the International Center for Elementary Particle Physics at the University of Tokyo. He is also a member of the Japan Academy.

## Personal Life

Masatoshi Koshiba resides in Tokyo, Japan, with his wife, Kyoko Kato, to whom he has been married since October 5, 1959. In addition to spending time with his children and grandchildren, he enjoys reading historical novels and listening to classical music.

## Further Reading

Gugliotta, Guy. "D.C. Scientist Shares Nobel in Physics; Three Are Honored for Space X-Ray, Neutrino Discoveries." *Washington Post* 9 Oct. 2002: A2. Print.

McFarling, Usha Lee. "3 Awarded Nobel Prize in Physics." *Los Angeles Times* 9 Oct. 2002: I10. Print.

Overbye, Dennis. "Nobels Awarded for Solving Longstanding Mysteries of the Cosmos." *New York Times* 9 Oct. 2002: A21. Print.

# Lang, Robert J.

## American physicist

**Born**: May 4, 1961; Dayton, Ohio

When Robert J. Lang quit engineering to pursue his other passion—origami, the Japanese art of paper folding—as a full-time career, he had written more than eighty technical papers and held forty-six patents on lasers and other optoelectronics (devices for emitting, modulating, transmitting, and sensing light). Now one of the premier origami artists in the world, Lang has created more than 480 designs and expanded the field both artistically and scientifically, finding practical applications for origami in fields ranging from medical technology to space exploration. Lang—whom Susan Orlean described for the *New Yorker* (February 19, 2007) as "composed, moderate, painstaking"—has been practicing origami since he was six years old, when a teacher gave him an instruction book to keep him from getting bored during math class. "The thing that got me hooked on origami all those years ago was the 'something for nothing' aspect," he told Chad Berndston for the South Boston, Massachusetts, *Patriot Ledger* (November 8, 2004). "The wonder of the things we can make has grown enormously." Among Lang's greatest artistic achievements is a full-size cuckoo clock, though in general he is best known for his figures that represent the natural world. Lang's shapes, which range from a simple banana slug to a koi fish complete with scales, can require more than three hundred folds; he is among an elite group of artists whose creations defy the normal scale of difficulty in origami, prompting the creation of the new "super-complex" category. "I've always loved problem-solving—it's part of being a scientist and an engineer," he told Berndston. "Coming up with a design is just another form of that, and while it's fun to fold an origami figure, it's also fun to solve the puzzle of how you create a new figure." That problem-solving also extends to the scientific realm, in which Lang's knowledge of origami has contributed to many advances, including ways to fold tricky items such as a heart stent, an airbag, and a telescope lens. Lang told Orlean that the field of origami is advancing rapidly, with almost limitless potential for both artistry and practical applications. "It's like math," he said. "It's just out there waiting to be discovered. The exciting stuff is the stuff where you don't even know how to begin."

## Early Life and Education

Robert James Lang was born on May 4, 1961, in Dayton, Ohio. His mother, Carolyn Lang, was a secretary and homemaker, and his father, Jim Lang, worked for the Airtemp division of the Chrysler Corporation. Lang has one older brother, Greg, who is a professor of horticulture at Michigan State University in East Lansing; his younger sister, Marla, is a commercial interior designer. The family moved frequently when Lang was young but settled in Atlanta, Georgia, by the time he was ten. Lang was always a "super-duper math whiz," his father told Orlean, recalling that his son often read the recreational math column in *Scientific American*. Lang became interested in origami at age six, when an elementary-school teacher gave him a book about origami to keep him entertained during math class. Those two interests—math and origami—persisted throughout the family's many relocations, with Lang creating his own origami figures from an early age. "I sought out all the books I could from the local library, and there weren't that many, and if I wanted to make an animal I couldn't find in the book, I wanted to try to make it up," he told Berndston. "All the people in the books had, so why couldn't I?" In high school Lang joined the school math team, which placed in state and regional competitions.

After graduating from high school in 1978, Lang enrolled at the California Institute of Technology (Caltech) in Pasadena. "Caltech was very hard, very intense," he told Orlean. "So I did more origami. It was a release from the pressure of school." The mathematical work involved in origami "revs your brain up into high gear," he explained to *Current Biography*, adding that the increased stimulation spilled over into other intellectual areas. "Origami was just a way of exercising your brain in a different direction from quantum mechanics," he said. Although he began to pursue origami seriously during that time, developing his own designs for a Pegasus, a parrot, a crab, and a squirrel, he kept his hobby largely to himself. "I guess I thought it was a kid's pastime that I hadn't grown out of," he told Orlean. "I was a little embarrassed about it." In 1982 Lang graduated from Caltech with a degree in electrical engineering and a growing stock of original origami designs. After earning a master's degree in electrical engineering at Stanford University in Palo Alto, California, Lang returned to Caltech to pursue a PhD in applied physics. While working on his dissertation, "Semiconductor Lasers: New Geometries and Spectral Properties," Lang created more than fifty origami designs, including those for a hermit crab, an ant, a skunk, and a mouse stuck in a mousetrap. From the beginning Lang's figures displayed not only sophisticated artistry but also striking realism. "They were dense and crisp and precise but also full of character: his mouse conveys something fundamentally mouse-ish, his ant has an essential ant-ness," Orlean wrote. For several months after receiving his PhD in 1986, Lang stayed at Caltech to finish his research; he then traveled to Germany for a ten-month postdoctoral program, studying semiconductor lasers. While in Germany, Lang also created an origami version of the traditional Black Forest cuckoo clock; described as a "tour de force" by Constance Ashmore Fairchild, writing for *Library Journal* (February 15, 2004), the figure took him three months to design and six hours to build.

## Life's Work

After completing his postdoctoral studies, Lang returned to Pasadena to work at the National Aeronautics and Space Administration (NASA) Jet Propulsion Laboratory, where he focused on optoelectronics. That same year, 1988, Lang also published his first collection of origami designs, *The Complete Book of Origami: Step-by-Step Instructions in Over 1000 Diagrams*. He followed that effort with *Origami Zoo: An Amazing Collection of Folded Paper Animals* (1990), written with Stephen Weiss, and *Origami Sea Life* (1990), cowritten by John Montroll. In 1992 he published his fourth book, *Origami Animals*, and became the first Westerner ever invited to address the annual meeting of the Nippon Origami Association in Japan. Meanwhile, Lang continued to work as an engineer, beginning a stint at Spectra Diode Labs in San Jose, California, in 1992.

In the early 1990s Lang began to integrate mathematics and origami, developing a computer program named Origami Simulation that
allowed the user to fold a sheet of paper on screen. When not working on such projects as fiber-optic networks for space satellites or the
patenting of self-collimated resonator lasers, Lang continued to pursue origami with an increasing number of side projects. In 1995 he

## "I've always loved problem-solving—it's part of being a scientist and an engineer."

published *Origami Insects and Their Kin: Step-by-Step Instructions in
Over 1500 Diagrams*, a collection of designs for twenty complex insect figures, and in 1997 he published *Origami in Action: Paper Toys
That Fly, Flap, Gobble, and Inflate*, a book of his collected designs
for beginners. In 1999 Lang was commissioned to create four works
for the Downtown Transit Mall in Santa Monica, California. He made
four animals representative of the area's fauna—a tree frog, a sea ur-
chin, a dragonfly, and a garibaldi (a species of fish)—which were then
cast in bronze and used to adorn drinking fountains around the city.
Lang's creations received the Southern California American Public
Works Association's award for streets and transportation project of the
year in 2002, and in 2005 Lang was commissioned to create two more
figures, a sea turtle and a flying fish.

Around that time Lang began to think seriously about quitting his
job as an engineer and pursuing origami full-time. Besides his work on
the Santa Monica project, he had been a consultant for companies or
government agencies that applied origami techniques in various ways.
The Lawrence Livermore National Laboratory, for example, hired him
to devise a way of folding a telescope that was to be sent into space.
Lang calculated that from those kinds of consulting jobs, as well as
his artistic commissions, he could make a living out of paper folding.
In addition to those projects, Lang wanted to create a different kind
of origami book. "Most origami books at the time were collections
of recipes for specific origami figures, but what I wanted to do was to
teach people how to come up with their own recipes," he explained to

*Current Biography.* That book, though, would require his full attention, he realized. The idea for the book came to him around the time of the dot-com bust, during which time the company he was working for, JDS Uniphase (which makes equipment used to build fiber-optic-telecom, data, and cable-television networks), lost a large portion of its business. By then Lang was in a managerial position at the company; with the bust, his duties largely changed from overseeing research and development to dealing with pay cuts and plant closings. "Laying people off was a lot less fun than inventing things," he told Orlean. "There were plenty of people doing lasers. The things I could do in origami—if I didn't do them, they wouldn't get done. Deciding to leave was a convergence of what I wanted to do plus what was happening at my company." Although he admitted to *Current Biography* that the idea of leaving a successful career in engineering for one in origami "sounds pretty crazy," he said that he was confident in his decision: "Sometimes, you know the crazy thing is the right thing to do. It was one of those few times when I knew it was the right decision."

The following year Lang published *Origami Design Secrets: Mathematical Methods for an Ancient Art* (2003)—which he refers to on his website as his "magnum opus"; he also published *Origami Insects II* (2003), a collection of eighteen new insect designs. That same year he was featured in the Origami Masterworks exhibit at the Mingei International Museum in San Diego, California. The exhibit, which included a six-foot-tall heron that Lang had folded, was originally supposed to run for six months, but it proved so popular that it was extended first by six months and then by an additional eight. Lang's pieces have also been featured in a number of commercial venues: he created a menagerie of toilet-paper origami animals as part of a commercial for the fabric deodorizer Febreze; a life-size model of Drew Carey for *The Drew Carey Show*; and airplane seats for the cover of an aircraft-seating magazine. He worked a week of fourteen-hour days to supervise the folding of hundreds of origami figures for a 2006 commercial for the Mitsubishi Endeavor, in which the car is driven through a world composed entirely of origami. Most recently, he completed a series of figures for a McDonald's commercial; the human figures, who wear such accessories as gold chains and Kangol hats, are folded from what appears to be McDonald's cheeseburger wrappers (actually origami paper printed with the McDonald's pattern).

Lang is constantly breaking ground in the area of practical applications of origami—specifically, fitting large items into small spaces. Among his designs are a pouch for medical instruments that can be opened without any unsterile surface touching a sterile one; a cell-phone antenna that fits inside the body of the phone; a heart implant that is folded into a tube for implantation, then unfurled when in place; and a computer model that simulates the folding and deployment of airbags.

Lang has also continued to develop two computer programs that allow for the mapping of complicated figures. TreeMaker, a program he began working on in the late 1980s and first released in 1994, can describe in mathematical terms any origami figure that can be represented by a stick-figure drawing—for example, an insect or a human, but not a cloud—and produce a replica of the creases necessary to fold the figure. Although some criticize the program as a perversion of the origami tradition, Lang said that he firmly believes that using a computer is an obvious step in the evolution of origami. "If people object to using computers to create origami, do they also object to using brushes to create a painting?" he said to *Current Biography*. Lang rarely uses TreeMaker in his work. When he created the program, he needed it specifically to make such complex features as deer's antlers and insects' legs. In general, however, he sees TreeMaker as a learning tool; by writing the program, he said, he gained valuable insight into the design process, which he could then translate into his more intuitive work. That intuitive work, he said, is what makes for the most amazing origami. "The computer is a tool. It's efficient for laying down a pattern or framework," he told Bennett Daviss for *New Scientist* (January 18, 2003). "But human skill will always be needed to implement the design. It's the person that makes the difference between creating something awkward and lifeless or something elegant and beautiful." TreeMaker is now on its fifth version; because it is freely distributed on Lang's website, he does not know exactly how many people have downloaded it, but he has estimated that a couple of hundred people do so every year.

While TreeMaker creates a pattern of creases, it cannot calculate the sequence in which the creases should be folded. To solve that problem, Lang created a different program, ReferenceFinder (currently in its fourth incarnation), which is also available for free download on his website. ReferenceFinder helps the origami artist find the most efficient folding pattern for a given crease; however, with upwards of two hundred creases per figure, it is still inefficient to run Reference-Finder on every crease. The artist must decide on his or her own which creases would be best to run through the program.

In his willingness to share origami designs and design strategy, Lang cofounded the Origami Design Challenge at OrigamiUSA's annual convention. Conceived by Lang and a friend, origami expert Satoshi Kamiya, the informal event involves no judges or winners—just the exchange of ideas. All of the contributors create the same general shape, but with different designs and, ultimately, different results. In 2004 Lang and Kamiya both brought versions of an origami beetle to the convention. The next year they announced that they would make hermit crabs, along with the eight other artists who had entered the challenge. In 2006 the ever-growing group made sailing ships. "It is a pleasant and remarkable situation, this mixture of sharing and competition that has remained nicely balanced for many years. Yes, we're all trying to outdo each other, but we're also 'giving away the store,' so to speak, by showing each other how our latest invention is constructed . . . ," Lang wrote for his website. "By this means, the entire art is advanced." In 2007 the competition will center on a complete plant—roots are optional, but the plant must have a stem. The emphasis will remain on the community of origami and on the personal challenge of pushing to do one's best, even without prizes. "The purpose of a design challenge is not rooted in some award, the joy of getting a bit of ribbon stuck onto one's effort," Lang wrote for his website. "The purpose of a design challenge is in the process, that it provides a spur and a goad to try something new and different. And the only opinion that matters is one's own. (Well, and maybe the opinions of other origami composers.)"

Lang is currently working on a life-size flying pteranodon for a permanent installation at the Redpath Museum at McGill University in

Montreal, Canada. He is one of the few Western columnists for *Origami Tanteidan Magazine*, the journal of the Japan Origami Academic Society. In January 2007 Lang was named editor in chief of the *Journal of Quantum Electronics*, published by the Institute of Electrical and Electronic Engineers, a position he held until 2010. He has also consulted for Cypress Semiconductor for several years.

## Personal Life

Lang lives in Alamo, California, with his wife, Diane, whom he met at Caltech as an undergraduate when they were both in a campus production of The Music Man. They have one teenage son, Peter.

## Further Reading

Kahn, Jennifer. "The Extreme Sport of Origami." *Discover Magazine* 29 July 2006. Print.

Kennedy, Louise. "Unfolding Origami's Secrets." *Boston Globe* 11 Nov. 2004. Print.

Orlean, Susan. "The Origami Lab." *The New Yorker* 19 Feb. 2007. Print.

# Levin, Janna

## American astrophysicist

**Born**: 1968; Texas

As a specialist in theoretical physics, astronomy, and cosmology, Janna Levin regularly grapples with big questions—so big that they are discussed in terms of light years. Levin, an assistant professor of astronomy and physics at Barnard College in New York City, uses mathematics to investigate characteristics of the universe, which is believed to stretch nearly 30 billion light years and to contain at least 50,000 billion billion stars—perhaps more than the number of grains of sand on all the beaches on Earth. (A light year is the distance that light travels in a single year, nearly six trillion miles.) Inextricably bound to questions about the size, structure, and evolution of the universe is another fundamental question: is the universe finite or infinite? If the universe is finite, a person in a rocket ship flying from Earth (hypothetically) in only one direction would approach Earth again billions of years later, just as a person walking (hypothetically) in one direction from a starting point in on Earth would eventually arrive back at the same location. With few exceptions, physicists and astronomers have long assumed that the universe is infinite, much the way people once believed that the world is flat. One reason for that assumption, at least among twentieth- and twenty-first-century scientists, is connected with the general theory of relativity, which was proposed by Albert Einstein in 1915. According to that extraordinarily revolutionary theory, the gravity of any star, such as the sun, warps space (and time); thus, space (or space-time, in the lingo of scientists) may be described as curved. That theory was corroborated by an experiment carried out during a solar eclipse in 1919, which showed, precisely as Einstein had predicted, that the gravity of the sun bends, or deflects, light rays from distant stars. The general theory of relativity led, mathematically, to the deduction that the universe is expanding, possibly without limit. Einstein himself, however, did not address the question of whether space is finite or infinite. Levin's research has led her to suggest that a finite universe is not only possible but probable; after all, as she has pointed out, scientists have never found anything in the natural world that appears to be infinite. "A finite universe, and indeed a finite universe with several extra dimensions, may be a prediction of a theory beyond Einstein's—the long coveted Theory of Everything," Levin wrote in a description of her research for her website. A physical theory of everything—another of Levin's research interests—is "the greatest ambition consuming theoretical physics," as she wrote. In the twentieth century, she continued, the mathematicians and logicians Kurt Godel and Alan Turing, and the mathematician and computer scientist Gregory Chaitin "proved that our knowledge of numbers themselves is fundamentally incomplete. . . . There are true relations among the numbers about which we can only prove that we can never prove them. Many times in the history of physics, theories have been shaped by such profound limits"—the limit of the speed of light, for example, and the limits of measurement described by the branch of physics known as quantum mechanics. "Alongside these should be listed the profound incompleteness in our knowledge of numbers—there can never be a mathematical theory of everything." Levin is seeking "to define the limits [that] mathematical incompleteness might set on a physical theory of everything. Just as Relativity emerged from the limit of light's speed and Quantum Theory emerged from the limits of measurement, deep insight into the universe and its origins could emerge by confronting the limit of mathematical incompleteness."

In an interview with David Kestenbaum for the National Public Radio program *Talk of the Nation* (July 12, 2002), Levin said, "Space is the whole story. And that's one of the really tough things to get over when you're first working on general relativity in Einstein's theory, and that is to accept that space is the whole thing. That is the whole universe. It is not expanding into anything. It's the whole story." In addition to her groundbreaking research in cosmology, Levin has written two critically acclaimed books, one a work of nonfiction and the other a novel of ideas. In *How the Universe Got Its Spots: A Diary of a Finite Time in a Finite Space* (2002), presented in the form of unsent letters to her mother, Levin wove descriptions of the cosmos and discussions of her research-related ideas with descriptions of her life in London as a postdoctoral fellow (1999–2003) and her reactions to various personal experiences. For her second book, *A Madman Dreams of Turing Machines: A Story of Coded Secrets and Psychotic Delusions, of Mathematics and War Told by a Physicist Obsessed by the Lives of Turing and Godel* (2006), she reimagined the lives of Kurt Godel and Alan Turing. Although the book was heavily based on biographies and other historical accounts, Levin and her publisher chose to classify it as a novel; the book jacket calls it both "completely imagined" and "entirely true." *A Madman Dreams of Turing Machines* brought Levin the 2007 PEN/Robert Bingham fellowship for writers, which honors "an exceptionally talented fiction writer whose debut work represents distinguished literary achievement and suggests great promise," according to the PEN American Center website. The novel also won the 2007 Media Ecology Association's Mary Shelley Award for outstanding fictional work and was a runner-up for the PEN/Hemingway Award.

During the 2003–04 academic year, Levin was the first scientist in residence at the Ruskin School of Drawing and Fine Art, as the department of fine art of Oxford University in England is known. "There were times when I thought that there's nothing I'd love more than to have become a painter, except I know that if I were a painter I'd be saying to someone, I wish I became a scientist, because I know that when I was exclusively focused on science I've said, I should have been a writer," she acknowledged on her website. Levin has maintained that in their quests to uncover truths, the worlds of science and art overlap. "We are in the questions we ask," Levin wrote in an article for *New Scientist* (August 19, 2006). "We are there, implicitly, in the search for meaning. And our stories do matter. Science without storytelling collapses to a set of equations or a ledger full of data. We are after more than that. For all of the respect paid to objectivity, science is ultimately a human endeavor embroiled in our complex themes. Stripping our discoveries of their narrative thread might lubricate scientific progress, but lending our discoveries to fiction gives our suppressed stories a chance to bloom."

## Early Life and Education

The youngest of three sisters, Janna Jhone Levin was born in Texas in 1968 to Richard and Sandy Levin and spent most of her childhood in Chicago, Illinois. She and her sisters were raised to believe that their gender did not limit their opportunities in life. Her father, a surgeon, "was always pointing out women of influence—doctors, politicians, athletes—with admiration. He was entirely disappointed if a girl opted for cheerleading over playing basketball, that sort of thing. My mother was definitely outspoken about women's issues and gender politics," Levin said, as quoted on her website. "It was a given that my sisters and I would go to college. It was a given we could do whatever we wanted. No one ever thought, let alone suggested, there was anything we couldn't do on the basis of our gender." (One of Levin's sisters is a veterinarian; the other is a lawyer.) During Levin's junior year of high school, her family moved to Florida, and she began to feel "restless," as she put it on her website. On her seventeenth birthday Levin and a friend of hers were injured when the friend's car tumbled off a footbridge and landed on its roof in a canal. The accident led her to realize that she no longer wanted to remain in her high school. Acting on the advice of one of her teachers, Levin left school before she earned a diploma and enrolled at Barnard College, an all-women's school associated with Columbia University in New York City. (The two schools are now formally linked.) At Barnard she intended to major in philosophy. An undergraduate course in astronomy led her to change her plans. As she recalled in *How the Universe Got Its Spots*, the professor, Joseph Patterson, "pierced my drowsiness with this one fact: Our bodies are made from elements synthesized in stars. These gloved hands of mine, these gloves, are reshaped atoms made only in the centers of stars. I admit I missed most of what he said before and most of what he said after, but I knew something had shifted for me."

In 1987 Levin won the Henry A. Boorse Prize in Physics, a Barnard award. She graduated from Barnard in 1988 with a BS degree in astronomy and physics. She then entered the Massachusetts Institute of Technology (MIT) in Cambridge, Massachusetts, where she earned a PhD in theoretical physics in 1993. Her dissertation was entitled "MAD Gravity and the Early Universe: A Possible New Resolution to the Horizon and Monopole Problems." In the abstract she wrote about the possibility that "the universe grows old while it is still hot. We call this era of prolonged aging the MAD era." As a postdoctoral fellow, Levin worked at the Canadian Institute for Theoretical Astrophysics in Toronto from 1993 to 1995, and at the Center for Particle Astrophysics at the University of California, Berkeley from 1995 to 1998. For four years beginning in 1999, she conducted research, wrote papers, and presented talks as an advanced fellow at the Department of Applied Mathematics and Theoretical Physics at Cambridge University in England. Her boyfriend (and later husband), Warren Malone, a bluegrass musician, handled domestic chores in their warehouse flat in London (a forty-five-minute train ride from Cambridge), where their neighbors included filmmakers and others active in the arts. Having friends who are artists and musicians "stretches me, so that I don't collapse in on myself," Levin told Jill Neimark for *Science & Spirit* (November/December 2002). "It's such a relief to be accepted for who I am rather than being seen as my work."

## Life's Work

In England Levin worked with the astrophysicist Joseph Silk, her one-time mentor at Berkeley, who had joined the faculty of Oxford University in 1999. She focused on several questions concerning the cosmos that Einstein did not address in his general theory of relativity. In addition to discussing the origin and expansion of the universe, Einstein determined that the gravity of stars and galaxies (aggregations of stars, gases, and dust, such as the Milky Way, that are bound by gravity) could warp space into different geometric configurations. The shapes of the warps—concave, flat, or hyperbolic—affect the paths in which light travels through the universe. But Einstein did not speculate about the overall shape or topology of the cosmos—that is, about whether the universe is infinite (an assumption widely held since 1576, when the English writer Thomas Digges offered that heretical idea) or is shaped in a way that makes it finite. Levin told David Kestenbaum that people preferred to think of the universe as infinite "because it seemed easier than adding new mathematics into [Einstein's] theory, but now that we're moving on—we're making some steps beyond Einstein's theory—we're starting to be able to realistically ask those questions." Levin believes that scientists may discover the shape of the cosmos by measuring the differences in temperatures in space since light began radiating through the universe, about 300,000 years after the big bang (the moment when, theoretically, the universe—including time

---

**"A finite universe, and indeed a finite universe with several extra dimensions, may be a prediction of a theory beyond Einstein's—the long coveted Theory of Everything."**

---

as well as space and matter—came into existence). While satellites have collected data showing that the temperatures of radiation are virtually identical in every direction, they have also located places where temperatures differ from the average temperatures by plus or minus 1/100,000th of a degree.

Levin and others have imagined a cosmos shaped something like a doughnut. In the 1970s the Russian physicist Dmitry Sokoloff tried to confirm that idea by scrutinizing images of distant galaxies in an attempt to find an image of Earth's galaxy, the Milky Way, at a moment in the past. (The reason an image of the Milky Way might be found in another part of the universe is that the light reaching Earth from the Milky Way originated tens of thousands or hundreds of thousands of light years in the past; because of the expansion of the universe, in the more-distant past the Milky Way was in a different location with respect to other galaxies.) Sokoloff never found an image of the Milky Way, but his inability to do so does not mean that such an image of it does not exist; rather, the image may be too faint to detect at present. If scientists ever conclude that the universe is finite, they will have to reconsider all the long-held theories that are based on the concept of an infinite universe that is expanding without limit.

Levin has also studied black holes, which she described on her website as "the most outrageous inhabitants of our universe, as well as the most elusive." A black hole is what remains when all of a star's fuel burns up; the star collapses into a sphere many magnitudes smaller than it had been. The sphere is extremely dense and exerts an extremely powerful gravitational pull. The force of its gravity is so great that even light becomes trapped inside what is known as the star's escape boundary (also called its event horizon). Although black holes are invisible to observers elsewhere in the universe, sometimes their effects on other stars can be detected. While she was working in England, Levin studied the possible effects of the collision of two black holes or of two black holes in orbit around each other. As she wrote on her website, "As black holes orbit each other, space itself ripples in response to the motion creating gravitational waves. If the black holes are rapidly spinning, then the orbit can be extremely irregular, even chaotic. Chaos imposes a fundamentally new perspective on the merger of black holes, gravitational wave detection, and possibly curved spacetime in general." Levin proposed that the calculations that scientists had been using to predict the effects of such cosmic events lacked the complexities that chaos would introduce. (In the case of such cosmic events, and many phenomena on Earth, "chaos" is a mathematical concept.) Levin's suggestion that chaos should be added to the standard calculations of such cosmic events elicited a defensive backlash from some eminent academics. "I got a pretty violent reaction," Levin told Anjana Ahuja for the *London Times* (February 11, 2002). "Some people . . . said it was important. Others flipped out and were just hysterical. They said there was clearly no chaos. I suppose they assumed that I thought they were being simple-minded; they felt I was disparaging their methods, which I certainly was not—they are all amazing people doing great work."

## Publications

Levin was living and working in England when she wrote her first book, *How the Universe Got Its Spots: A Diary of Finite Time in a Finite Space* (2002). "I had been writing a lot of technical articles, and I was so tired of writing in such a formal way, and it felt like I was reaching such a small audience," she told Kestenbaum, referring to her motivations for committing to paper what might be described as both a memoir and a popular-science book about cosmology. The book contains descriptions of her research and philosophical ruminations about the universe, along with accounts of her sometimes isolating experiences as a scientist and often difficult relationship with her boyfriend. Although a few critics complained that Levin had failed to strike a satisfactory balance between her discussions of cosmic events and her musings on personal concerns, most found it to be both intellectually and emotionally engaging. "This intimate account of the life and thought of a physicist is one of the nicest scientific books I have ever read—personal and honest, clear and informative, entertaining and difficult to put down," Alejandro Gangui wrote for *American Scientist* (September/October 2002)."[Levin's] female perspective is refreshing, and her personal account is firmly aimed at non-experts, . . . yet avoids patronizing readers," Valerie Jamieson wrote for the *Times Higher Education Supplement* (October 4, 2002). *Discover Magazine* named *How the Universe Got Its Spots* one of the twenty best science books of the year.

After completing her research at Cambridge, Levin spent nearly a year as the first scientist in residence at the Ruskin School of Drawing and Fine Art in Oxford, supported by a Dream Time fellowship from the National Endowment for Science, Technology, and the Arts, a British funding group. In that capacity Levin, who since childhood has loved to draw and paint, wrote pieces for and about London artists and musicians, including catalog essays and book chapters. She also began writing what became her next book, the novel *A Madman Dreams of Turing Machines*, about two of the most brilliant, original, and influential thinkers of the twentieth century: Kurt Godel and Alan Turing, both of whom suffered greatly in their personal lives, Godel because of severe mental illness and Turing because of Great Britain's intolerance of homosexuality. (British authorities punished what they called his "indecent acts"; they were unaware that Turing, in top-secret work, had cracked the Nazis' Enigma Code for the British in World War II.) Levin's book contains alternating scenes in which each man appears without the other. Godel and Turing never met, but their professional and private lives had much in common—some of Turing's work grew out of Godel's theories—and both died tragically: Godel starved himself to death in the paranoid fear that he would be poisoned, and Turing committed suicide by eating a cyanide-laced apple. A third character in the novel is the nameless physicist who narrates the story and ponders the ideas and experiences of Godel and Turing. As a way of organizing her ideas while writing, Levin sketched out scenes and other parts of the book into a kind of storyboard, which she would redraw and reorganize as her thoughts evolved.

"Levin turns dry history into a literate and revealing examination of the power and the peril of creative genius," Laurence Marschall wrote for *Natural History* (July/August 2007), in one of many admiring reviews of *A Madman Dreams of Turing Machines*. In another, John Allemang wrote for the Toronto *Globe and Mail* (September 21, 2006), "Though her book comes with endnotes on historical accuracy, this isn't a case of a scientist being nervous around her new genre. Levin makes the transition seem effortless, as if equations and adjectives were instruments in the same pursuit." "The ultimate achievement of this book is not in its creative biographical studies, but in its exploration of the sociology of ideas," Harold Heft wrote for the *Montreal Gazette* (December 23, 2006), while Mark Sarvas, in a critique for the *Philadelphia Inquirer* (October 1, 2006), wrote, "Levin's novel is, at its heart, an effective meditation on interconnectedness, language and free will . . . a paean to misunderstood heroes, as well as an elegy to a lost golden area of discovery." Sarvas and Ariel Swartley, writing for the *Los Angeles Times* (August 21, 2006), both found Levin's handling of her narrator less than felicitous, while a reviewer for *Publishers Weekly* (August 25, 2006), in one of the few unenthusiastic assessments of the book, wrote, "Levin is sympathetic to all concerned, but doesn't quite make a larger point, dramatic or otherwise."

In 2004 Levin returned to New York City, to become an assistant professor of astronomy and astrophysics at Barnard College. As a prominent woman in male-dominated fields, Levin has frequently been asked to comment on her experiences in academia. Levin told Ahuja that science departments were "the last bastion of maleness" and that "the whole system is structured under the assumption that you have a housewife and your wife's life doesn't matter." In an essay for *Newsweek* (September 25, 2006) entitled "This Topic Annoys Me," she wrote, "I don't ever want to talk about being a woman scientist again. . . . I was never very good at telling those stories because truthfully I never found them interesting. What I do find interesting is the origin of the universe, the shape of space-time, and the nature of black holes. . . . Why should curing sexism be yet another terrible burden on every female scientist? After all, I don't study sociology or political theory. I study the history of the cosmos written in the laws of physics. That's a story I tell pretty well."

In 2003 Levin won a Kilby International Award, "created to honor unsung heroes and heroines who make significant contributions to Society through Science, Technology, Innovation, Invention, and Education," according to the website of the Texas-based foundation that supports that prize. She took a leave of absence from Barnard during the spring 2008 semester. In March 2008 Levin launched her podcast, titled *Janna Levin Reports from the Cosmos*. Each installment features discussions of topics in cosmology, among them black holes, the big bang, and the nature of the universe. The podcast is available through Levin's website. In 2012, Levin was awarded a fellowship by the John Guggenheim Memorial Foundation.

Levin and her husband live in New York City with their two young children.

## Further Reading

Battersby, Stephen. "Opinion Interview: Janna Levin, Theoretical Cosmologist." *New Scientist* 6 Apr. 2002: 40. Print.

"Dr. Janna Levin Discusses Her Idea of a Finite Universe." *Talk of the Nation/Science Friday*. Natl. Public Radio, 12 July 2002. Web. 8 Aug. 2012.

"Jonathan Lethem and Janna Levin." *Seed Magazine*. Seed Media Group, 5 Mar. 2007. Web. 8 Aug. 2012.

Overbye, Dennis. "Universe as Doughnut: New Data, New Debate." *New York Times* 11 Mar. 2003: F1. Print.

# Mansfield, Peter

## British physicist

**Born**: October 9, 1933; London, England

Peter Mansfield was informed that he had been awarded the Nobel Prize for physiology or medicine three days before his seventieth birthday. "The phone rang at about 10:15 in the morning and they'd told me I'd won. I just didn't believe it," he told Andy Done-Johnson for the *Nottingham Evening Post* (October 7, 2003). "Then I got a second call from Stockholm and it started to sink in. There was always the hope that I might get to this stage, but all you can really do is plod on." The award was a huge surprise, he told John Crace for the *London Guardian* (October 14, 2003). "There were no whispers at all on the academic grapevine. Ten years ago there was [some] talk [concerning Nobel Prizes] and I suppose there was an air of expectancy around that time. But when nothing happened, we rather assumed the moment had passed, and forgot about it." Amid the surprise and celebration, however, Mansfield informed Crace, there was also a little regret: "I started work on MRI back in 1972, and it does feel a little strange to get the award thirty years later. You also can't help wondering how much more you might have achieved had the recognition come earlier. But that's the way things have worked out." Nevertheless, Mansfield told Nigel Hawkes for the *London Times* (October 7, 2003), "I am grateful they [the Nobel committee] have looked to us here in Nottingham. This is a great honour for me, a great honour for British science in general and especially for the people who are continuing to work in this area at the University of Nottingham."

### Early Life and Education

The physicist Peter Mansfield was born on October 9, 1933, in London, England, to a working-class family. The son of a gas fitter, Mansfield lived through World War II and the Nazi Blitz of London, developing a fascination with the German-made V1 and V2 rockets that terrorized the city as the conflict neared its end. "I knew all about rockets from the wrong end," Mansfield informed Sean Kirby for the *Nottingham Evening Post* (January 1, 2001). "I was eleven years old towards the end of the war when the V2s came in. It was a dreadful time, you couldn't hear the rockets coming. But like all the boys at that time I was interested, I used to collect shrapnel." After attending Cork Street Primary School, Mansfield went on to the William Penn School, located in Peckham, in southeast London. At fifteen, however, the future Nobel Prize winner left William Penn—though he insists that he did not drop out: "When I was eleven," Mansfield explained to a reporter for the *Nottingham Evening Post* (October 13, 2003), "I went to what they used to call a [central] school, which was half-way between a grammar school and a secondary modern. They had a sixth form and I would have happily continued my education. But at the end of my first year they did away with [central] schools and we were all packed off to a secondary modern—where you were expected to leave school at fifteen, and no arguments." The school guidance counselor informed the departing young man that he had no future in science; consequently, Mansfield began working as a compositor in a London printing house while preparing for his O-levels at night school. (O-levels are British academic examinations usually taken at about age sixteen.) Over the next three years, Mansfield worked in the print shop but explored his interest in explosives and rocketry in his spare time. This dedication paid off when, at eighteen, Mansfield left the printing industry and London for a position in Aylesbury, Buckinghamshire, as a scientific assistant at the Rocket Propulsion Department (RPD), a division of the British Ministry of Supply. Mansfield's tenure in the RPD was cut short, however, when he was conscripted into the army. Given his expertise in explosives and rocketry, Mansfield assumed the army would assign him a posting where these skills would be utilized; however, "It was square pegs and round holes," Mansfield told Kirby. "I already knew quite a bit about weapons and explosives. All that was mentioned and I wanted to go into an engineers or ordnance unit—so of course I was put into the service corps as a storeman." Despite the undesirable assignment, Mansfield put in his time, and when he was demobilized he began studying for his A-levels (British examinations that gauge preparedness for university-level study).

Mansfield was accepted into Queen Mary College (now Queen Mary and Westfield College), part of the University of London, and obtained his bachelor of science degree in 1959. Having had his academic career interrupted by work and the army, Mansfield, in comparison with his fellow students, took his studies at the university very seriously: "I knew I was there to work," Mansfield explained to Crace. "And because of that I did better than most of those who had had a supposedly better education." He added, "Many students now consider it their right to have a car, or go out drinking most nights. I'm not teetotal, but I knew what I could afford, and throughout my time at university I probably only drank three half-pints." As an undergraduate Mansfield began his research into nuclear magnetic resonance (NMR)—the name was later changed to magnetic resonance imaging (MRI) after the word "nuclear" took on negative and misleading connotations. "My project," Mansfield informed Kirby, "was to build a portable NMR detector which could be carried around for architectural research, detecting objects underground." Mansfield developed such a device and his findings were reported in the *New Scientist* magazine. In 1962, three years after completing his bachelor's degree, Mansfield received his PhD in physics from Queen Mary College, having done his thesis on the use of NMR in the study of chemical structure—not biological structure, in which the technology would eventually find its most effective use and for which Mansfield would receive his Nobel Prize.

## Life's Work

After earning his PhD, Mansfield traveled to the United States to work as a research associate in the physics department of the University of Illinois. Two years later, in 1964, he returned to England to become a lecturer in physics at the University of Nottingham, where he has remained ever since, becoming a senior lecturer in 1968, a reader in 1970, and a full professor in 1979. Mansfield's decades-long tenure at Nottingham was interrupted only for a year-long sabbatical (1972–1973), during which he taught as a visiting professor at the Max Planck Institute in Heidelberg, West Germany.

On October 6, 2003, the Karolinska Institute awarded the Nobel Prize in Physiology or Medicine to both Peter Mansfield and Paul C.

Lauterbur, a professor at the University of Illinois, "for their discoveries concerning magnetic resonance imaging." MRI is a noninvasive technique that enables doctors to see inside the human body; it

---

**"I am grateful they [the Nobel committee] have looked to us here in Nottingham. This is a great honour for me, a great honour for British science in general and especially for the people who are continuing to work in this area at the University of Nottingham."**

---

is consequently an extremely valuable diagnostic tool. The procedure is conducted more than sixty million times each year and is widely regarded as one of the transformational innovations in medicine of the last half century. In a press release published on the Nobel e-Museum website, the Nobel Assembly explained the properties of magnetic fields that make MRI possible: "Atomic nuclei in a strong magnetic field rotate with a frequency that is dependent on the strength of the magnetic field. Their energy can be increased if they absorb radio waves with the same frequency (resonance). When the atomic nuclei return to their previous energy level, radio waves are emitted." From these radio waves, researchers can then extrapolate data regarding the molecules that sent out the signals. "In effect," Hawkes wrote, "the nuclei of atoms are being used as tiny radio transmitters to broadcast information about themselves."

## Magnetic Resonance Imaging

MRI scanners have become so effective in medical diagnosis because approximately two-thirds of the human body is made up of water. "When a patient lies down in an MRI scanner, an intense but harmless magnetic field makes [protons] in the water line up like so many microscopic compass needles," Ian Sample explained in the *London Guardian* (October 7, 2003). "The scanner then uses pulses to give the [protons] a gentle flick. MRI scanners measure how the [protons] respond and use that to piece together a picture of the organs and tissues."

The full potential of MRI took decades to be realized. The first major discoveries were made in the 1940s by Felix Bloch and Edward Mills Purcell, who found they could identify a substance by bouncing a radio wave off it and analyzing the frequency that resulted. (Bloch and Purcell were jointly awarded the Nobel Prize in Physics in 1952.) In the early 1970s, Paul Lauterbur developed a method to determine a signal's origin "by creating a small difference in the magnet between [the] right and left side . . . so that he could tell by just the pitch of the signal where the signal was coming from," Elias Zerhouni, the director of the National Institutes of Health, explained in an interview for the *NewsHour* with Jim Lehrer (October 6, 2003). By knowing the origin of such signals, researchers could now map out the underlying structure. Converting the signals into meaningful pictures, however, proved a time-consuming process.

In 1972, using magnetic gradients, Mansfield demonstrated that the signals from the scanner could be analyzed using mathematics and then converted into images. Mansfield recalled to John Crace, "We experimented with imaging layers of camphor and material coated in Vaseline. By rotating the sample we could clearly see the five layers of the plate structure. But it was soon apparent that these artificial systems were too academic." So the researchers applied the technology to analyzing such biological structures as plants. Eventually Mansfield used MRI to make an image of a student's finger; he published the picture in a 1976 article for the *British Journal of Radiography*. With this first application of MRI to humans, the technology began to gain acceptance in the medical community. In the years that followed, MRI equipment was introduced into hospitals; today there are approximately 22,000 MRI machines in use worldwide. The machines have proved very effective for diagnosing such ailments as back pain, multiple sclerosis, and cancer. Mansfield has also pioneered the practice of fast gradient variations, or echo-planar scanning, which greatly accelerated the MRI imaging process and enabled doctors to analyze real-time changes in brain chemistry; this process is known as functional magnetic resonance imaging, or FMRI.

Besides the Nobel Prize Mansfield has been the recipient of numerous awards and accolades, including the Gold Medal (1983) of the Society of Magnetic Resonance in Medicine; the Gold Medal and Prize (1984) from the Royal Society Wellcome Foundation; the Duddell Prize (1988) from the Institute of Physics; the Silvanus Thompson Medal (1988) from the British Institute of Radiology; and the Royal Society's Mullard Medal (1990). In June 1986 Mansfield was elected a fellow of Queen Mary College. In February 1987 he was elected a fellow of the Royal Society. From 1987 to 1988 he served as president of the Society of Magnetic Resonance in Medicine. In the 1993 New Year's Honours, Mansfield was knighted by Queen Elizabeth II. He was awarded honorary membership in the British Institute of Radiology in 1993, and is an honorary fellow of the Royal College of Radiology. He received an honorary doctorate from the University of Strasbourg, in 1995; from the University of Kent, at Canterbury, in 1996; and from the Jagellonian University, in Krakow, Poland, in June 2000.

## Personal Life

Mansfield has been married to his wife, Jean, for more than forty years. They have two daughters, Sarah Crosbie and Gillian Haywood. Sarah lives next door to her parents and says of her father, "He's such a wonderful man and we're very proud of him. He's always lived for his work and even now he's retired he's still hard at it. . . . This [the Nobel Prize] is wonderful and he really deserves it," Done-Johnson reported. In addition to his research, Mansfield enjoys languages and reading and was an experienced pilot of airplanes and helicopters before health concerns precluded his flights. After his climb out of poverty to scientific acclaim and knighthood, Mansfield reportedly remains humble and unpretentious, as his daughter remarked to Done-Johnson: "He owns a Rolls-Royce which he hardly ever takes out because he's far happier pottering about in his little Ford Ka [a relatively inexpensive model available in the UK]."

Though he retired from teaching in the mid-1990s, Mansfield continues to conduct his research, laboring at Nottingham and his Magnetic Resonance Center five days a week. He is currently working with his own new company, General Magnetic, to build MRI scanners that are not as loud as current models and thus pose no threat to the ears of children and the unborn.

## Further Reading

Judson, Horace Feeland. "No Nobel Prize for Whining." *New York Times* 20 Oct. 2003. Print.

Lawrence, Felicity. "Pater Mansfield, the Doctor Under Scrutiny." *The Guardian.* Guardian News and Media Limited, 7 Aug. 2001. Web. 9 Aug. 2012.

Mansfield, Peter. "Autobiography." *NobelPrize.org.* Nobel Foundation, Oct. 2003. Web. 9 Aug. 2012.

"Physicist Shares Nobel Prize for Medicine." *PhysicsWorld.com.* Institute of Physics, 6 Oct. 2003. Web. 9 Aug. 2012.

# McKay, Christopher P.

## American astrophysicist

**Born**: c. 1954; Florida

"Life can be in places and do things you can't imagine," the planetary scientist Christopher P. McKay said to Michael Tennesen for *Discover* (July 1989). One of McKay's goals is the formulation of a precise definition of life, one that "can guide a search for life outside Earth," as he wrote for *PLoS Biology* (September 14, 2004). "With only one example of life—life on Earth—it is not all that surprising that we do not have a fundamental understanding of what life is," he noted. "We don't know which features of Earth life are essential and which are just accidents of history." McKay launched his career at the Space Science Division of the Ames Research Center, an arm of the National Aeronautics and Space Administration (NASA), in 1982, the year he earned his doctoral degree, and he has worked there ever since. His specialties are astrobiology, the study of life in the universe, and astrogeophysics, the study of the physical processes that occur on planets other than Earth and those of the solar system's many moons. His main research interests include the origins of life on Earth and how it evolved; the possible existence of life on Mars when the solar system was much younger; and the properties of highly unusual microbial species called extremophiles, which flourish in earthly environments that are extraordinarily harsh and would quickly kill other creatures: places that are extremely cold, hot, dry, acidic, alkaline, salty, or radioactive, for example, or that appear to be devoid of nourishment.

The air on Mars, the fourth planet from the sun, is mostly carbon dioxide, and it is less than 1 percent as dense as the air on Earth; in addition, for much of the Martian year, the surface of the planet is frozen, its temperature far below zero. Such an environment would not be conducive to survival for the vast majority of living organisms found on Earth. Mars does not appear to harbor living organisms, but the discovery a few decades ago of extremophiles has led scientists to hypothesize that billions of years ago, the red planet, as it is called, might have hosted organisms impervious to conditions that are intolerably harsh for most forms of life on Earth. To understand how extremophiles survive, and how similarly hardy organisms on Mars might have survived, McKay has studied various extremophile species in their Mars-like natural habitats, including the driest desert on Earth (the Atacama in Chile) and the coldest places on Earth (parts of Siberia in Russia, the Canadian Arctic, and Antarctica). McKay is a leading proponent of a process known as terraforming: altering an environment, in this case that of Mars, so that it can sustain life. McKay believes that warming Mars and changing the density and composition of its air could restore its biosphere. "Mars lived fast, died young, and left a beautiful body," he said during the seven-part "Great Terraforming Debate," posted on *Astrobiology Magazine* (June 14, 2004), a NASA publication. "We could . . . just ignore it, or we could do something better and bring it back to life. Mars is beautiful the way it is, but I think it would be even better if we could restore the biosphere that it once had. . . . In other words, give Mars back its heartbeat." McKay also said in that debate, "I'm not proposing to send life from Earth there. That's only the last resort. If Mars has no genome, then we could share ours with it. But I personally think that a Mars full of Martians is much more interesting than Mars full of Earthlings."

McKay is also an expert on Titan, the largest of the many moons of the planet Saturn, and he has worked with scientists on the building of, and interpretation of data from, the Cassini-Huygens probe, which passed through Titan's atmosphere in early 2005. For the past five years, he has participated in NASA's Robotic Lunar Exploration Program, which aims to put astronauts on Earth's moon by 2020. He is also a co-investigator on two NASA projects: the first, the Phoenix Mars Mission, landed an instrument-laden robotic spacecraft on the surface of Mars in 2008 and, three months later, found evidence of frozen water there. The second, the Mars Science Laboratory Rover Mission, was launched on November 26, 2011. The rover will explore its surroundings to determine whether the environment was once hospitable to microbial life. "I'm primarily interested in this work because I think one of the primary questions is, 'Are we alone?,'" McKay told

Rosanne Spector for the *Chicago Tribune* (July 7, 1991). "And for me, 'we' is anything living. . . . We're all related to the same primordial mother."

## Early Life and Education

Christopher P. McKay was born in about 1954 and raised in Florida. He has recalled playing in orange groves with his sister and seven brothers. His interest in science dates from his childhood. "I got a chemistry set one Christmas, and I read a lot of science books," he told Pamela S. Turner for her book *Life on Earth—and Beyond: An Astrobiologist's Quest* (2008). He was also mechanically inclined; after he acquired his own motorcycle, he completely disassembled it, removing every bolt and even taking apart each gasket, and then reconstructed it. McKay told an interviewer for Spiked-Online.com that the late-1960s TV series *Star Trek* inspired him, because it propounded the idea that "science could lead to a positive future and that there were great things still to do—great voyages of discovery waiting to be made. I was determined to help make that future real, and to participate in those great discoveries. It was all too easy to be disenchanted with science, at a time when the future looked grim and science was often blamed. Science had given us atomic weapons, chemical pollution, and dwindling ozone. Science was not the hero of the bright new future, it was the villain. But *Star Trek* imagined that the future could be bright—that science and technology could be used to create a positive and humane future." In high school McKay enrolled in every science course available to him, and physics proved to be his favorite. "In physics there were just a few basic concepts you could apply over and over again," he told Turner. "But I hated marine biology. There were too many facts you had to memorize, like the name of this or that squid."

After high school McKay attended Florida Atlantic University (FAU), which has campuses in Boca Raton, Fort Lauderdale, and five other locations. He majored in physics and mechanical engineering. By chance he found a neglected telescope in a laboratory closet; after he repaired it, he took up stargazing. While at FAU, using instructions in books, he built his own telescope. "I was amazed you could take sand and two pieces of glass and make all the lenses you need for a telescope," he recalled to Turner. McKay earned a bachelor's degree in 1976.

That year McKay enrolled in the graduate program in astrophysics at the University of Colorado (CU) at Boulder, a school prominent in that field. The year 1976 also marked the end of the voyage of the NASA spacecraft *Viking 2* to Mars. The orbiters of that craft and its sister craft, *Viking 1* (which had arrived at its destination in 1975), sent back to Earth thousands of photographs of Mars's surface, and they transmitted to Earth the information gathered by the two *Viking* landers, whose instruments examined materials on the surface of Mars and gases in its air. The photos showed canyons, volcanoes, and craters, while the equipment on the landers detected molecules of carbon dioxide ($CO_2$,) water ($H_2O$), and nitrogen ($N_2$) in the atmosphere but no evidence of organic molecules (which, on Earth, contain carbon-hydrogen bonds and are present in all living organisms). "The results were most mysterious to me," McKay said, as quoted on a NASA website. "Here was a planet with all the elements needed to support life ($CO_2$, $H_2O$, $N_2$) present in its atmosphere, with evidence of liquid water in the past and yet there was no sign of life. It seems like Mars had 'the lights on but nobody home.' I slowly became more and more interested in life and how it originates, survives, and changes a planet." He has traced his fascination with astrobiology to the *Vikings'* discoveries.

McKay soon met other CU-Boulder students who shared his interest in the Martian landscape and atmosphere. They agreed that NASA should consider terraforming Mars and sending astronauts to the planet to explore it firsthand, and they decided to petition CU-Boulder administrators to offer a course to examine that idea. Inspired by the book *Project Icarus* (1968), whose authors, students at the Massachusetts Institute of Technology (MIT), described a class at MIT devoted to the feasibility of deflecting an asteroid on a collision course with Earth, McKay spearheaded the initiative and helped to organize the course. "Due to bureaucratic issues, which I no longer remember, I was actually listed as the instructor of this class," he said during an interview posted on the website of the Mars Society, "and at the first class meeting we had more than twenty students. In addition to terraforming Mars we also considered human exploration as a step toward terraforming." Some of the students—whom a journalist dubbed the "Mars underground"—organized a conference to discuss the possibility of sending humans to Mars. Called "The Case for Mars" and held in 1981, the conference was a great success. "We were stunned so many people came," McKay told Turner. "We had real scientists come, people from NASA." The book *Case for Mars*

---

**"The moon was never alive. Mars is dead. The question in my mind is—should we bring it back to life? Well, I vote for life."**

---

(1984), edited by Penelope J. Boston, contains papers presented at the conference. In the next decade and a half, five more "Case for Mars" conferences were held.

Earlier, in the summer of 1980, McKay had worked as a NASA planetary-biology intern at the Ames Research Center at Moffett Field in Sunnyvale, California. He was guided by James B. "Jim" Pollack, a world-renowned planetary scientist, and became acquainted with Imre Friedmann, a Florida State University biologist who also worked at Ames and whose interests included astrobiology and extremophiles. Through Friedmann, McKay joined a team of scientists associated with Florida State University's Cryptoendolithic Microbial Ecology Research Project. The scientists were studying microbes in the Dry Valleys of Antarctica, a region with unusual biological, chemical, and physical characteristics. McKay's job was to build sensors that monitored rock formations' temperature, light, moisture, and other features. His work involving microbes that flourished in an environment usually described as hostile to life led him to think about the possibility that living organisms might live or have lived in an environment even more extreme: that of Mars. When the internship ended McKay returned to CU-Boulder and completed his dissertation. Entitled "Photochemical-Thermal Model of Hydrogen and Oxygen in the Summer Mesosphere and Implications for Noctilucent Cloud Formation," it was about the formation of a type of cloud that has been detected only in the mesosphere (the third of the five layers of Earth's atmosphere), about 30 to 50 miles above Earth's surface. McKay earned a PhD in astrogeophysics in 1982, and that year he joined the Ames Center as a full-time research scientist.

## Life's Work

In collaboration with the biologist Robert Wharton, McKay began to study the mechanisms through which microorganisms survive in frozen Arctic lakes; working with Wharton, he became a research diver for NASA's Antarctic Lakes Project. One of the project's main goals was to identify the chemical "signatures" left by dead microorganisms embedded in sediment (particles of matter that have been carried by water and then deposited), the longer-term aim of which was to develop a means for future NASA explorations of Mars to find evidence of life there—not existing life, as had been attempted earlier, but extinct forms of life. From 1990 to 1991 McKay was a member of a team looking for frozen bacteria in Siberia. The scientists dug 150 feet into the arctic permafrost—permanently frozen soil, sediment, and rock—and removed chunks of material to test for life. They discovered bacteria that could be revived after having remained in a frozen state for three million years. "It may be possible that bacteria frozen into the permafrost at the Martian south pole may be viable," McKay later said, as quoted by Vincent Kiernan in *New Scientist* (March 5, 1994).

In an article titled "Making Mars Habitable," published in *Nature* (August 8, 1991), McKay, Owen B. Toon, and James F. Kasting discussed the possibility of terraforming the red planet. The scientists suggested that "planetary engineering" could alter Mars's environment to make it hospitable to humans. Paraphrasing their argument, Robert Cooke wrote for *Newsday* (August 8, 1991), "Terraforming the planet breaks down naturally into two steps: warming the planet, and altering its chemical state. Warming could be linked to an artificial greenhouse effect. If the temperature of the Martian surface were increased, say, by warming the poles with giant mirrors, or by spreading black soot over the polar caps, or by introducing greenhouse gases, then the amount of carbon dioxide and water vapor in the atmosphere would also increase."

McKay was a member of the National Geographic Society's expedition to the Gobi Desert of Mongolia from 1991 to 1992 and of the NASA/United States Geographical Survey Expedition into the Lechuguilla Cave in Carlsbad Caverns National Park in New Mexico from 1994 to 1995. The cave—the fifth-longest on Earth and the deepest in the continental United States—is of particular interest because, in the absence of any outside sources of energy, virtually no plants or fungi live there (except near the entrance); the bacteria that live in the cave thrive on sulfur, iron, and manganese—three of the ten most abundant elements in Martian soil. "On Mars, if life has survived it has to be deep underground," McKay told Eduardo Montes for the Associated Press (April 18, 1994). McKay joined the McGill University expedition to Axel Heiberg Island in the Canadian Arctic (1995–1996), and the 1994, 1995, and 1997 NASA expeditions to the Atacama Desert. Because of its extreme dryness, the Atacama Desert resembles Mars more closely than any other locality on Earth. On parts of the Atacama, no rain has ever been recorded; parts of it harbor virtually no life, and in others, the scant flora and fauna—most species of which live nowhere else in the world—exhibit highly unusual survival mechanisms. In 2005, by means of a mobile robot of the sort that could be used on Mars, McKay and his coworkers detected microbes in the soil of Atacama.

In 1998, at Pavilion Lake in British Columbia, Canada, McKay and another NASA team studied rocklike mounds known as microbialites: assemblages of microbes and sedimentary particles and chemicals. Most of the few microbialites that have been discovered on Earth are fossilized forms of the widespread accretions of once-living organisms and inorganic materials that thrived on Earth one billion to two billion years ago. By studying the Pavilion Lake's modern-day analogues of those ancient specimens, scientists may be able to figure out ways to recognize microbial fossils on Mars. McKay and seven of his colleagues described their studies of Pavilion Lake microbialites in *Nature* (October 5, 2000).

In 2003, at a meeting of the American Astronomical Society's Division for Planetary Sciences, McKay presented a paper in which he proposed that billions of years ago, a "second genesis" had occurred on

Mars, in which life forms whose proteins, DNA, and genes were vastly different from those that evolved on Earth had developed. In 2004 he participated in the Astrobiology Science Conference, organized by *Astrobiology Magazine*, the Seattle SciFi Museum, and Breakpoint Media. During the conference he discussed the ethical implications of altering Mars, taking the view that humans have an obligation to restore Mars to its formerly hospitable state—even if terraforming the planet for human habitation is unlikely to happen anytime soon. "Human beings are a particular subset of life that require particular conditions," McKay said, as quoted by *Space Daily* (June 18, 2004). "And it turns out oxygen in particular is very hard to make on Mars. That is, I think, beyond our technological horizons—it's a long time in the future. But warming Mars up, and restoring its thick carbon dioxide atmosphere, restoring its habitable state, is possible." Arguing that Mars was once a living planet, McKay said, "We see that distinction for the first time when we look beyond the Earth, when we look at the moon. There's nature; there's no life. When we look at Mars, we also see nature, probably no life. It's different from the moon, and we lack the word that distinguishes between something that's dead, and something that was never alive. The moon was never alive. Mars is dead. The question in my mind is—should we bring it back to life? Well, I vote for life." McKay also advocates allowing any life found on Mars to exist without human interference. He told Jim Gilchrist for the *Scotsman* (August 3, 2001), "We should be intellectually prepared to face this issue and to defer to the indigenous Martians—however microscopic—and even assist them in regaining biological control of their planet. We will be the better for this, in both ethical and scientific terms." McKay has written or co-written articles about ethical issues regarding Mars for publications including *Scientific American* (December 1990), the *Journal of the Irish Colleges of Physicians and Surgeons* (Volume 22, Number 1, 1993), the *Planetary Report* (July/August 2001), and the *International Journal of Astrobiology* (2011); his essay "Does Mars Have Rights? An Approach to the Environmental Ethics of Planetary Engineering" is included in the book *Moral Expertise: Studies in Practical and Professional Ethics* (1990), edited by Don MacNiven.

McKay has been involved with research regarding Earth's moon since 2004, when NASA announced that it had assigned management of its Robotic Lunar Exploration Program to the Ames Research Center. Components of the program are the design and construction of a new spaceship, a voyage of astronauts to the moon by 2020, and the construction of a NASA base on the moon. "An exploration science program with a sustained human presence on the moon gives us the opportunity to conduct fundamental science in lunar geology, history of the solar system, physics, and the biological response to partial (Earth) gravity," McKay told *Space Daily* (November 21, 2005).

McKay has proposed sending seeds to Mars and Earth's moon to test plants' responses to conditions on those bodies. A robotic lander would transport seeds from the plant *Arabidopsis thaliana* (a member of the mustard family, which includes cabbages and radishes, that is often used in plant research) in a growth module (a miniature greenhouse). Soil from the Martian or lunar surface would then be added, and instruments in the module would monitor the plants' progress. The seeds would have been genetically modified so that the plants would glow in different colors if they were suffering in particular ways, to alert those monitoring them of trouble. McKay and the University of Florida plant biologist Robert Ferl presented those proposals at a meeting of lunar scientists held at the Ames Center in July 2008. McKay told Richard Macery for the Sydney (Australia) *Morning Herald* (April 6, 2002) that a plant growing on the moon or on Mars would be both "a true biological pioneer" and a "powerful symbol of the long-term vision of life expanding beyond the Earth."

McKay has edited or co-edited several books, including *Comets and the Origin and Evolution of Life* (2006). He sits on the board of the nonprofit Planetary Society, whose members include astronauts, scientists, engineers, and lay space enthusiasts. The society's mission is to advance the human understanding of space. McKay serves on the US Committee for the International Permafrost Association and as an assistant editor of the journal *Planetary and Space Science*. He has won the Kuiper Award from the Division of Planetary Sciences of the American Astronomical Society, as well as the Ames Honor Award and the Group Achievement Award, both from NASA.

## Further Reading

Booth, William. "Why Not Implant Life on Mars?" *Washington Post* 17 Dec. 1990. Print.

"Giving Mars Back Its Heartbeat. Great Terraforming Debate: Part I." *Space Daily.* SpaceDaily.com, 16 June 2004. Web. 8 Aug. 2012.

"Red Dragon Mission: A Discussion with Chris McKay." *The Mars Quarterly* 4.1 (Summer 2012): 4–5. Print.

Turner, Pamela S. Life on Earth--and Beyond: An Astrobiologist's Quest. Charlesbridge, 2008. Print.

# Myers, Joel N.

## American meteorologist

**Born**: November 3, 1939; Philadelphia,
Pennsylvania

Not so long ago, people relied on old sayings to predict the weather—
"Red sky in morning, sailor take warning," for example. These days,
professional forecasters have Doppler radar and satellite images at their
disposal, and predicting the weather has become a profitable business.
Joel N. Myers, the founder and CEO of AccuWeather, is one of the
leaders in the industry. Myers started his company in 1962, as a graduate student at Pennsylvania State University, at a time when weather
forecasting in the United States was primarily the domain of the federal government. He has revolutionized the forecasting industry by
providing accurate and localized weather forecasts to businesses and
government agencies that require more than generalized reports; today
his company serves more than 15,000 clients worldwide, including
more than one-third of Fortune 500 companies. AccuWeather's four
hundred employees, among them one hundred full-time meteorologists, also supply weather forecasts to domestic and international media outlets, including more than 750 newspapers, 250 radio stations,
and 200 local television stations, as well as such major channels as
CNN. In 1994 AccuWeather began providing free weather reports on
its website, AccuWeather.com; as of June 2004 the site was drawing
some 3.4 million unique users a month, according to Dave Gussow in
the St. Petersburg *Times* (August 13, 2004). In addition, the company
supplies weather reports to major news websites, among them those
of the *New York Times* and the *Washington Post*. Recently Myers's
company pioneered the "RealFeel" temperature gauge, which some
industry analysts believe will ultimately replace the wind-chill factor
as a more accurate representation of how cold outside air feels.

Although countless people disparage weather forecasters for having spotty records when it comes to accuracy of predictions, the

science of weather forecasting improved greatly during the second half of the twentieth century, since the development of mathematical models and the first computer-aided weather forecasts in the 1950s. Today meteorologists say that the five-day forecast is about as reliable as the three-day forecast was fifteen years ago. AccuWeather faces heavy competition—from the Weather Channel, which has a powerful presence through its cable channel and the Internet; from the more than one hundred small firms that, like AccuWeather, offer specialized forecasts for paying clients; and from the National Weather Service, the federal agency that produces most of the raw data that is used and interpreted by the other commercial weather companies. Despite the presence of such competitors, AccuWeather enjoys a strong reputation and continues to grow. For Myers, who spent years teaching meteorology at Pennsylvania State University, weather forecasting is an art as much as it is a science. "Warren Buffet beats the S&P virtually every year," he told Fred Guterl for *Newsweek* (September 30, 2002), referring to the noted investor Warren Buffet's ability to produce higher returns than the Standard & Poor 500 stock-market index. "It's the same thing with weather forecasters. You have some who can and some who can't. It's a matter of skill and talent."

### Early Life and Education

Joel Norman Myers was born on November 3, 1939, in Philadelphia, Pennsylvania, to Martin Henry Myers and Doris A. Schwartz Myers. His fascination with the weather began at an early age; by five or six he was already keeping a daily weather diary. "I remember being awestruck around the age of five when it snowed," Myers recalled to William Ecenbarger for the *Chicago Tribune* (January 11, 1987). "When I was eleven, I concluded that the weather affected every facet of life and got the idea to sell forecasts. Then on Oct. 15, 1954, Hurricane Hazel hit. That was something. I was in a constant state of excitement for thirty-six hours. We had a barometer [an instrument that measures atmospheric pressure] at Central High School, and I watched it drop all day long. I called the weather bureau every hour. While I was delivering my newspapers, I saw a screen door blow off a house. That evening I stood on the front porch with my father and brother, watching the wind and the trees bending to touch the house. I was mesmerized by the weather!"

In his early teens Myers received as a gift from his father a twenty-year-old barograph (a recording barometer) that cost about $60—a considerable sum in the early 1950s. "It was a big deal to get," Myers recalled in an article edited by Kimbra Cutlip for *Weatherwise* (July/

---

### "Long-range forecasting is witchcraft."

---

August 2001). "It had an eight-day clock, and I ran it regularly for years." At about the same time, Myers set up his own small-scale station to measure and record the weather. After he graduated from high school, he entered Pennsylvania State University (Penn State), then one of the few schools in the country with a meteorology program. He earned a BA degree in the field in 1962, after which he entered the university's graduate program in the same discipline.

## Life's Work

Myers was spurred into starting his forecasting business in 1962, when a local gas utility company contacted Penn State during a search for a meteorologist to predict the weather for the approaching winter, which would allow the firm to gauge likely consumer demand. The head of the meteorology department referred them to Myers. In November 1962 Myers founded one of the first commercial weather-forecasting companies in the country. (According to some reports, Myers, lacking an office at first, had to report his earliest forecasts from a phone booth.) The company acquired its present name, AccuWeather, in 1971.

For a while Myers had trouble persuading businesses to purchase forecasts from him when they could obtain free forecasts from the radio and television reports delivered by the National Weather Service (NWS). The NWS is an agency of the US government that monitors climatological data and issues daily weather forecasts and severe-weather warnings. It traces its roots back to 1870, when President Ulysses S. Grant authorized military stations to relay storm warnings;

the Department of Agriculture assumed control of the growing operation in 1890. Today the NWS uses complex computer systems to amalgamate and report on information received from satellite images and Doppler radar as well as a network of more than 10,000 volunteers across the country who make daily weather measurements, monitoring such variables as temperature, humidity, wind speed, and barometric pressure.

Myers secured his first one hundred clients by phoning some 25,000 businesses and giving them free trials of the AccuWeather service. He carved a niche for himself by providing not only more-accurate forecasts than the NWS, but detailed, localized reports as well. His first customers included a Penn State football coach and area ski resorts. The latter would prove to be among Myers's more important early clients, as they depended heavily on accurate reports so as to anticipate when they would need to use their snow-making equipment.

As AccuWeather began to build a customer base, Myers continued to earn his credentials as a meteorologist. He received a master's degree and PhD in meteorology from Penn State in 1963 and 1971, respectively, and worked as a weather broadcaster on television and radio between 1962 and 1985. In addition, he taught at Penn State between 1965 and 1981, as an assistant professor of meteorology.

In 1971 Myers decided to broaden the sale of AccuWeather's forecasts to media outlets. In a successful strategy, he got one television and one radio station in each major American metropolitan market to buy his forecasts; then, just as he had hoped, the stations started heavily promoting their "exclusive" AccuWeather forecasts to distinguish themselves from their competitors. Myers's company thus gained free publicity and, before long, additional customers. By 1983 AccuWeather's client list had swelled to five hundred and included some forty television stations as well as 120 radio stations and many newspapers. Also in 1983 AccuWeather began operating internationally, providing tailored forecasts for key crop-growing areas around the world.

For AccuWeather clients not connected with the media, specific weather predictions are important for various reasons. Investors and corporations, for example, refer to weather forecasts when deciding to buy or sell products, particularly crops and commodities. If a speculator knows that storms at sea will affect the transport of oil to the United States or other parts of the world, for instance, he or she can then surmise that the price of oil will go up. In 1977 Accu-Weather's forecasters predicted a cold front that threatened to bring freezing temperatures to the citrus-producing regions of Florida. The forecasters pooled $5,000 of their own money and that of their fellow office workers and invested in frozen orange juice, assuming correctly that the cold front would prove devastating to that year's citrus crop. In two weeks their investment generated $40,000 in earnings for them. Other clients who depend on AccuWeather's predictions include school-district administrators, in anticipating school closings in the event of snowstorms; city-management officials, in order to prepare for snow removal; trucking companies, in planning transportation routes; construction companies, when laying concrete or erecting skyscrapers; and department stores, because weather can significantly affect consumer behavior.

In AccuWeather's early days, Myers was fond of telling reporters, "Long-range forecasting is witchcraft"—an assessment with which those in the industry would readily agree at that time. As computer models, radar, and satellite-imaging technologies have improved, however, AccuWeather's team of meteorologists have grown confident enough to say that the current five-day forecast is as accurate as the three-day forecast had been in previous decades. In addition, Myers told Fred Guterl that AccuWeather meteorologists are able to make fifteen-day forecasts "with some skill," although predicting the movements of storms is notoriously difficult. In the late 1990s Accu-Weather began offering custom forecasts for businesses up to a month in advance. Though not as reliable as five-day forecasts, those predictions have become a necessity for many companies, which often pay up to $15,000 a month for AccuWeather's services.

By 1987 there were about one hundred private firms offering the kinds of weather-forecasting services AccuWeather had pioneered. AccuWeather, like its rivals, relies heavily on raw data from the NWS, which it then filters through its own computer models and interprets for its clients. Many weather models are available, and accuracy depends upon using the right one. But just as important, meteorologists must be able to look beyond computer models and study weather patterns everywhere on Earth for developments that could affect conditions in North America several weeks in the future. "The thing is, a lot of meteorologists are lazy," Myers told Guterl. "They rely too much on the models. Very few look far beyond the United States." AccuWeather scientists examine data not only from the NWS but also from the World Meteorological Organisation and national weather services in Europe, Japan, and Canada. In 1999 AccuWeather's meteorologists astutely relied on data from Europe during Hurricane Floyd, which struck the Bahamas and southeastern United States, allowing them to predict the storm's trajectory better than the Weather Channel.

In 1994 Myers established AccuWeather.com, an online service providing free streaming-video forecasts, weather maps, and related information. (Paying subscribers receive access to additional features.) The site is popular, as Myers explained to Karen Brown for *Broadband Weekly* (June 4, 2001), because, for example, "if you are traveling to San Francisco or New York or whatever, what you can do is get on and get the five-day forecast with a talking head presenting it . . . just like you would see the weather in those cities."

Though the company faces more competition than ever before, AccuWeather has maintained a solid customer base, with a customer-renewal rate of 98 percent, according to Myers. In addition, although Weather.com, the website of the Weather Channel, is ranked as the most popular single source of information about weather on the Internet, AccuWeather provides material to more than 1,200 other sites, which in total, according to the AccuWeather Web site, draw an estimated 70 million unique users each month.

Myers is forging ahead with new weather forecasting technologies, including the patent-pending "RealFeel" temperature gauge, which the company introduced in 2000. In the past meteorologists have tried to present a more accurate evocation of an individual's experience of temperature by calculating for such factors as wind chill and humidity. The RealFeel formula, by contrast, takes into account multiple parameters, including temperature, wind, humidity, sunshine intensity, cloud cover, time of day, precipitation, and elevation. Myers, who spent years developing the formula with other meteorologists at

AccuWeather, has said that once RealFeel is patented, he will publish the system for other scientists and weather researchers to study. The company's other technological innovations include a state-of-the-art meteorological database, begun in 1979, and OnLine with AccuWeather, an award-winning educational tool. The company has also developed a service for wireless customers that sends streaming video directly to cell phones and other wireless devices.

## Personal Life

Joel N. Myers has three children: Daniel Martin, Sharon Annetta, and Erika Ann. He is a member of the American Meteorological Society, Sigma Xi, Phi Kappa Phi, Rho Tau Sigma, and Sigma Gamma Epsilon. He has written many articles and has appeared on such television shows as *Larry King Live*. Since 1981 he has been a trustee of Pennsylvania State University. In 2004 he received the National Weather Association's Operational Achievement Individual Award for his pioneering efforts in the field of meteorology.

Myers's leisure-time activities include collecting antique barometers. According to the *Weatherwise* article, Myers has acquired more than 220 barometers, most of which still work; they line the walls of his home and office and range from one-and-a-half-inch-diameter barometers from the early 1900s to ships' barometers to an angle barometer from 1774. Banjo barometers, so-named because they resemble the size and shape of a banjo, are among his favorites; some "that have real large faces . . . 15 inches across are just magnificent," he told *Weatherwise*. "The combination of the wood and the instrumentation, that's what makes them a thing of beauty."

## Further Reading

Astor, Dave "Four Decades for Forecasting Firm: Newspaper List Growing as Accu-Weather Marks Anniversary." *Editor & Publisher* 23 June 2003: 21. Print.

Ecenbarger, William. "Divining the Skies." *Chicago Tribune* 11 Jan. 1987: V1. Print.

Herrera, Stephan. "Weather Wise." *Forbes* 14 June 1999: 90. Print.

# Nakamura, Shuji

## Japanese electronics engineer and inventor

**Born**: May 22, 1954; Seto, Japan

In 1993 a virtually unknown Japanese researcher at an obscure company announced the successful creation of a product that to those outside the world of materials science might sound equally humble: a light-emitting diode (LED) that shined bright blue. Yet the announcement, made about six months after the discovery, startled scientists around the world. Few people knew of Shuji Nakamura, the person who, working virtually by himself, created what has often been referred to as the holy grail of LEDs, or of his company, then called Nichia Chemical Industries. Moreover, scientists studying LEDs generally thought that the compound gallium nitride (GaN) was probably the least promising of materials on which to base a blue LED, but Nakamura's success hinged on his use of it. Between 1994 and 1999 Nakamura and Nichia rapidly brought to market almost every major innovation suggested by his initial discovery—leaving almost every other company in the world trailing behind. In addition to creating the first extremely bright, truly green LED, Nakamura fashioned a blue laser that was small and reliable enough to be used in consumer electronics, preparing the way for the blue-violet laser used by the HD-DVD and Blu-ray digital-storage formats, which were released in mid-2006 with an eye toward replacing the DVD. Nakamura also developed, in 1995, a white LED, a technology that many experts predict will someday take the place of the filament light bulb. Since LEDs can theoretically be almost perfectly efficient producers of light—not wasting most of the energy put into them in the form of heat, as filament bulbs do—Nakamura's invention could also significantly reduce the amount of energy needed to light the planet. In his first book, *The Blue Laser Diode: GaN Based Light Emitters and Lasers* (1997), coauthored with Gerhard Fasol, Nakamura wrote: "What I have managed to achieve shows that anybody with relatively little experience in the field, no big money and no collaborations with universities or other companies, can achieve considerable success alone when he tries a new research area without being obsessed by conventional ideas and knowledge." At the end of 1999 Nakamura came to the United States to accept a professorship at the University of California, Santa Barbara. In 2006 he received the second-ever Millennium Technology Prize, a more than $1.2 million award created and funded by the Finnish government and described on its website as the "world's largest technology prize." Pekka Tarjanne, a Millennium Prize Foundation representative, called Nakamura "a splendid example of perseverance and dedicated research work," according to the organization's website. "He has worked with great determination for decades, and even severe setbacks have not prevented him from achieving something that other workers in the field regarded as almost impossible: using a reactor system of his own design to develop a solid material, in this case gallium nitride, into a powerful light source producing blue, green and white light, and also creating a blue laser." A writer for *ScienceWatch* (January-February 2000) suggested, "Nakamura may have single-handedly, or virtually so, changed the technological face of the world."

## Early Life and Education

Shuji Nakamura was born on May 22, 1954, in Seto, a tiny village on the coast of Japan's smallest island, Shikoku. Little information is currently available in English about Nakamura's early life, but Bob Johnstone, in the *London Independent* (October 10, 1999), described Nakamura's father as a maintenance engineer for a local power company; other press reports suggest that Nakamura's interest in science was spurred in part by the comic book and television series *Astro Boy*, especially Astro Boy's kind-hearted guardian, Professor Ochanomizu Hakase. Drawn to math during his high-school years, Nakamura was persuaded by a teacher that engineering offered him a more secure future and that he could follow his interests freely after passing the necessary entrance examinations and beginning at a university. Yet once Nakamura started at Tokushima University, also on Shikoku, he discovered that "the reality was completely different," according to his 2001 book *Break Through with Anger*, as translated and excerpted in a July 2001 article by Chiaki Kitada in Japan Inc. (online). "I was so shocked and disappointed that I felt like I was falling into hell from heaven. I asked myself, what have I been doing for the past eighteen years?" Locked into his chosen course of study, Nakamura went on to earn three degrees in electrical engineering from Tokushima: a bachelor's in 1977, a master's in 1979, and in 1994, long after leaving his formal study, a doctorate in recognition of his work on GaN and blue LEDs.

## Life's Work

After receiving his master's degree, Nakamura turned down an offer for a job at Kyocera, a major electronics company, in part because he was put off by the crowds and expense of metropolitan Japan. By then a husband and father, Nakamura chose to live in the Tokushima Prefecture and work for Nichia Chemical. Located in the town of Anan, Nichia had been established in 1956 by Nobuo Ogawa, a chemist who started out selling highly purified calcium but went on to manufacture phosphors—light-emitting chemical compounds used to coat light bulbs and the tubes that create images on some televisions and computer monitors, among other applications. Nichia was small, with only about two hundred employees at that time, and little known outside its circle of industrial clients. It was unusual among Japanese firms in never setting aside a regular budget for research, but allocating resources as needed. "Ever since we started, we've always had what Japanese people call a twisted navel, a hesomagari," Eiji Ogawa, then the president of Nichia, told Bob Johnstone for *Wired* (March 1995). Having a "twisted navel," Ogawa continued, "means that we are a bit perverse—we don't copy what other people do, we use a slightly different method."

When Nakamura joined Nichia, he was the company's lone electrical engineer and the only scientist among its small research-and-development staff to hold a master's degree. His group was already working on altering the metal gallium and, over the next three years, Nakamura helped produce a crystal form of gallium phosphide (GaP) that drew electricity under certain circumstances and resisted it under others, a property that made it a semiconductor. Since the material

Nakamura created could be used to produce visible red and yellow-green light when a current moved through it in a single direction, Nakamura's crystal could be cut into very thin sheets, known as wafers, to form some of the most important ingredients in an LED.

Nichia's slight research budget did not provide Nakamura with a staff or equipment to help with the development of the material, so he had to undertake all aspects of the work, including assembling his own equipment. The technical challenge of, for example, building a reactor able to withstand the 2,700 degrees Fahrenheit needed to form the crystal forced him to expand his understanding of the production of LEDs far beyond his university training. It also regularly put him in danger, since once every other week or so materials in his handmade reactor would explode. Nakamura told Glenn Zorpette for *Scientific American* (July 5, 2000) that the first few times it happened, "people would come in asking, 'What happened? Nakamura, you still alive?'" Eventually, his colleagues became inured to explosions, as did he. "I thought my life [would] be like this forever," he told Zorpette.

By the time Nakamura's GaP crystals reached the market in 1982, larger and better-known companies than Nichia were already offering them, bringing Nichia's sales of the substance to only about $10,000 a month. "My company wasn't happy with me," Nakamura told *ScienceWatch*. He then turned to developing a commercially viable gallium arsenide (GaAs) crystal for LEDs and in 1985 succeeded—only to have it fail in the marketplace. Changing his approach slightly, Nakamura, now the head of one of the company's research-and-development groups, spent the next three years working to produce wafers for red and infrared LEDs. Nakamura's aluminum gallium arsenide epitaxial wafers "were just as good" as the competition's, he told *ScienceWatch*, and "the prices were the same, but our company was small and local and couldn't compete. So once again my company was not happy."

**Developing Blue LEDs**

In early 1988, frustrated by continually having his products fail because of their commercial redundancy, Nakamura decided he wanted to try a much riskier venture: creating a blue LED. The impetus was straightforward. No academic or commercial research group seemed close to making the type of blue LEDs that would have broad application, yet their potential value was obvious. Blue LEDs would not only expand the colors produced by LEDs—in itself valuable for making full-color displays—but white LEDs could be made by adding this newly fashioned color to the already existing greens and reds. Moreover, a material that emitted light with the kind of mixed optical qualities of wavelength, phase, and direction of an LED—a light that was, in other words, incoherent—should also be able, with modifications, to emit coherent light, making it a laser. With its comparatively short wavelength, a blue laser could be honed to read (and write) much smaller points on a compact disc-like material, allowing engineers to increase the number of points and thus the amount of data that could be stored there. For consumer electronic applications, such a laser had to be extremely bright but also operate at room temperature, using only modest amounts of energy, for at least 10,000 hours.

Nakamura has often said that when he first approached his immediate supervisor with this idea, he was rebuffed. "All the big companies and universities haven't been able to do that," Nakamura recalled his manager saying, as he related the conversation to Zorpette. "Why do you think you can do it at a small company?" Still determined to go forward, Nakamura then approached Nobuo and Eiji Ogawa—an unusually bold decision in a highly hierarchical corporate environment. "I was very mad," Nakamura told Zorpette. "I wanted to quit Nichia. I didn't care about anything. It was OK for them to fire me. I was not afraid of anything."

Nakamura earned surprisingly strong support from the Ogawas. Nakamura was "good at thinking," Nobuo Ogawa told Bob Johnstone for the *Independent*. "So my policy was: if I let him get on with it, he'll probably be able to come up with the goods." Nakamura has also suggested that his track record, however mixed in some respects, gave Nichia's executives confidence in his abilities. "They had faith in me because, despite the dismal sales, I had developed three new products for this company and I was the only one at Nichia who succeeded in making new products," he told *ScienceWatch*.

Eventually the company authorized an unusually high budget of about $3.3 million for research and development, approximately 1.5

percent of Nichia's annual gross sales revenue. The Ogawas also allowed Nakamura to spend a year in the United States, working as a visiting research associate at the University of Florida in Gainesville, where he hoped to learn about metal-organic chemical vapor deposi-

---

**"What I have managed to achieve shows that anybody with relatively little experience in the field, no big money and no collaborations with universities or other companies, can achieve considerable success alone when he tries a new research area without being obsessed by conventional ideas and knowledge."**

---

tion (MOCVD), then an emerging technology for producing the wafers needed for more unusual LEDs.

Not long after arriving in Florida in early 1988, Nakamura found out that the MOCVD machines he had hoped to work on were no longer available to him, so he spent the next ten months assembling one himself. Though the process gave Nakamura a detailed understanding of how MOCVD machines function—an insight that proved crucial to his creation of blue LEDs—he felt slighted by the other scientists around him, who considered a person with a master's degree and no history of publications to be "an engineer, not a researcher," as Nakamura told Zorpette. Returning to Japan in 1989, he resolved to begin publishing papers on his research in order to bolster his international reputation and to follow the American model of pursuing his individual research agenda, regardless of what the company demanded. "I ignored every order coming from my boss and stopped answering phones, attending meetings, or helping the sales staff," Nakamura wrote in *Break Through with Anger*.

Nakamura also decided to concentrate his efforts at producing a blue LED by using GaN. Though GaN had been shown to produce a faint blue light two decades before, most researchers looking into making blue LEDs saw greater promise in zinc selenide (ZnSe). Nakamura chose the more unusual material, he told *ScienceWatch*, after thinking "about my past experience: if there's a lot of competition, I cannot win. . . . Even if I succeeded in making a blue LED using zinc selenide, I would lose out to the competition when it came to selling it."

Nakamura developed his blue LEDs over the next three years, working at times against the explicit orders of Eiji Ogawa and, defying company policy, publishing papers on his discoveries. One important element in his success was his modification of a commercially built MOCVD machine that added an extra stream of gas flowing perpendicular to the material on which he was making the wafer. This allowed him to eliminate problems in the GaN caused from heating it to over 1,800 degrees Fahrenheit. Using this system Nakamura could fine-tune the compound as needed.

One evening in March 1992, Nakamura wrote in *Break Through with Anger*, "my blue LED shined quite unexpectedly." It continued emitting light for about forty-eight hours but was still too faint to be used commercially. In January 1993 Nakamura drew from his experimental LED a deep blue light that achieved the brightness of one candela. (A standard international unit of measurement, the candela has less than 1 percent of the brightness of a standard 100-watt bulb. Nakamura's first significantly bright LED was nonetheless 100 times brighter than other LEDs existing at the time.) Nichia waited until November 1993 before announcing Nakamura's discovery, however, because it wanted to be able to sell the product at the same time. US scientists, who first saw a selection of Nakamura's blue LEDs at a conference in April 1994, met them with noisy appreciation, and, Robert F. Service wrote in *Science* (April 22, 1994), "amidst the murmurs were undoubtedly a few groans of envy."

Suddenly a prominent scientist, Nakamura was made a senior researcher at Nichia in 1993. An article detailing his discoveries, co-authored by Takashi Mukai and Masayuki Senoh, was published in 1994 in the professional journal *Applied Physics Letters* and proved highly influential, being cited by other researchers 572 times between 1994 and early 2000, according to *ScienceWatch*. Another 1994 paper, published in the journal of the *Japanese Applied Physics Society*, won the society's award for best paper that year; he won the award again in 1997.

Between 1989, when he published his first paper at the relatively advanced age of about thirty-five, and the end of 1999, Nakamura published nearly two hundred scholarly essays, as well as his first book. Nakamura is now listed by the Institute for Scientific Information as a "highly cited" researcher, a designation given to "less than one-half of one percent of all publishing researchers—truly an extraordinary accomplishment," according to the institute's website. During that same decade-long span, Nakamura also pushed Nichia to seek patents for his work, something the company had generally considered risky, since anyone reading the patent could learn important details about the manufacturing process. Still, Nichia carried through and, after leaving the company, Nakamura estimated that the company had sought patents on about five hundred processes and products he had developed during that time; Nichia was eventually awarded eighty patents in Japan and ten in the United States. An experienced patent reader himself, Nakamura had worked to ensure that the patents left few ways for anyone outside Nichia to use the underlying technologies. "Any form of blue LED or violet laser would infringe these patents," Nakamura told Yoshiko Hara for *Electronic Engineering Times* (April 24, 2000). After he left the company and tried to reclaim control over most of the rights to his inventions, Nakamura discovered how successful he had been in helping tie the patents securely to Nichia alone. "I regret it now," Nakamura told Hara. "At the time I had no intention of quitting. . . . Now being outside the company, I know it's very difficult to bypass the patents. It's almost impossible."

Nakamura's final five years with Nichia proved exceptionally productive. In 1995 Nakamura fashioned a green LED that was both much brighter than most others and truly green (previous attempts had tended toward yellow). That same year he increased the brightness of his own blue LEDs and later announced that, by coating the diode at the heart of his blue LED with a phosphor, he made a white LED. During this same time, he and his development team at Nichia were working to create a blue laser from his GaN semiconductor material. At the beginning of 1996 Nakamura announced that he had been successful, and in

October he again surprised the rest of the laser research community by presenting a blue laser that emitted as a continuous wave—a requirement for typical consumer electronic uses—rather than pulsing the way his earlier model had. Even this more advanced version, however, operated for only thirty-five hours, far short of the 10,000-hour minimum Nakamura needed to achieve; he reached that level at the end of the following year. Then Nichia announced that it would begin making violet lasers commercially available in October 1999, a time when the company's competitors, Yoshiko Hara claimed, "boasted that theirs emitted for a while before dying."

Around the same time that Nichia made its announcement about violet lasers, Nakamura publicly acknowledged that he was leaving the company. Paid only about $80,000 per year when, by his calculation, his inventions had quadrupled the company's sales from roughly $100 million to $400 million, Nakamura at first planned to enter the American private sector but soon realized that doing so would leave him and his employer vulnerable to suits from Nichia. Instead, he decided to accept a professorial position at an American university. After being approached by a number of high-profile universities, Nakamura settled on the University of California, Santa Barbara, where he joined a group of six other professors and some thirty graduate students dedicated to further exploring the potential of GaN. Formally joining the faculty in 2000, Nakamura was made the Cree Chair in Solid State Lighting and Display the following year, soon after the chair was funded by the North Carolina-based semiconductor company Cree.

Nakamura had begun consulting with Cree, one of Nichia's most important competitors, in early 2000, and in December of that year Nichia announced that it was suing Cree and Nakamura for infringing on the company's patents. The following year, Nakamura retaliated by bringing a suit against his former employer, seeking 80 percent of the rights to the most important of the blue LED patents and his fair share of the profits, which he calculated to be about $16 million. According to press reports, Nakamura had received only $160 for each of his many inventions for Nichia, even the blue LED, which he argued would bring Nichia $1.4 billion in revenue over the life of the patent. While acknowledging that the suit was directly prompted by

Nichia's own, Nakamura emphasized that he was fighting in part to change Japanese corporate culture, which too rigidly placed teamwork above individual innovation, stultifying creativity. "Japan owes much of its prosperity to its talented engineers," Nakamura told Irene M. Kunii for *BusinessWeek* (December 10, 2001). "But no matter how great their contribution, few receive compensation that matches those of their senior managers."

Over the next three years Nakamura's case and his criticisms of Japanese society prompted widespread debate in Japan and abroad. Nakamura gave public lectures and appeared on television-news shows and even advertisements, making at times biting criticisms of his home country. Moving to the United States, as Nakamura told Yuri Kageyama for the Associated Press (December 12, 2001), made him feel "as though I defected from a communist country to a free country. Everyone [in the United States] is given an equal chance to pursue the American dream. In Japan, everyone gets to be the eternal salaryman." In September 2002 a judge in Tokyo ruled that Nichia was, unconditionally, the rightful patent holder, eliminating the possibility that Nakamura might be able to recover any control over the blue LED technology. In January 2004 a district court in Tokyo ruled that Nichia owed him about $185 million in compensation for his inventions. According to the *Japan Economic Newswire* (January 30, 2004), judge Ryoichi Mimura justified the large award because the blue LED "was a totally rare example of a world-class invention achieved by the inventor's individual ability and unique ideas in a poor research environment at a small company." The award was significantly reduced, however, by the time the case ended in early 2005, when a higher court forced Nakamura and Nichia into making an out-of-court settlement for about $7 million. "The judicial system in Japan is rotten," Nakamura said when the decision was announced, according to a report by Yuri Kageyama for the Associated Press (March 8, 2006). "I am outraged." Later Nakamura did acknowledge that his case had made companies rethink their compensation policy for engineers. In that sense, he told Norimitsu Onishi for the *International Herald Tribune* (May 2, 2005), "I'm very much satisfied by the results that the lawsuit has brought about."

In addition to the Millennium Prize announced in June 2006, Nakamura has received multiple awards from the Institute of Electrical and Electronics Engineers, including a 1996 prize for engineering achievement, the 1998 Jack A. Morton Award for his contribution to solid-state devices, and two 2001 awards, one for distinguished lecturer and one for quantum electronics. In 2002 alone he received a World Technology Award for information technology, given by the World Technology Network; an Economist Innovation Award from the Economist magazine in the category "no boundaries"; and the Benjamin Franklin Medal in engineering from the Franklin Institute. That year he also received 50 percent of the Takeda Foundation's second-ever award for a technical achievement enriching wider social and economic well-being. Nakamura has also received the Prince of Asturias Award for Technical Scientific Research (2008) and the Harvey Prize for advancements in technology and science (2009).

A book by Bob Johnstone on Nakamura's achievements and the implications of his work, *Brilliant!: Shuji Nakamura and the Revolution in Lighting Technology*, was published in 2007.

**Personal Life**

Nakamura and his wife, Hiroko, have three daughters: Hitomi, Fumie, and Arisa. They live in Southern California. Nakamura has said he plans to return to Japan only once he fully retires.

**Further Reading**

Hara, Yoshiko "Blue Laser is Born of Bright Spirit." *Electronic Engineering Times* 30 Oct. 1997: 262. Print.

Hara, Yoshiko. "Laser, LED Developer Dreams in Color." *Electronic Engineering Times* 24 Apr. 2000: 189. Print.

Johnstone, Bob. "True Boo-Roo." *Wired Magazine* 3.03 (Mar. 1995). Print.

Zorpette, Glenn. "Profile: Inventor of the Blue-Light Laser and LED, Shuji Nakamura." *Scientific American* Aug. 2000: 30–31. Print.

# Rahmstorf, Stefan

### German oceanographer and climatologist

**Born**: February 22, 1960; Karlsruhe, Germany

Along with all but a tiny fraction of his colleagues worldwide, the climatologist Stefan Rahmstorf is certain that Earth's climate is warming. The dangerous rates at which the temperatures of the oceans and the atmosphere are rising can be traced, most scientists believe, directly to the activities of humans since the start of the Industrial Revolution in the 1700s—and primarily to the growing use of fossil fuels. A native of Germany, Rahmstorf is a physicist and an oceanographer as well as a climatologist; in Potsdam, Germany, he holds the title of professor of physics of the oceans at the University of Potsdam, and he heads the department devoted to the analysis of Earth systems at the Potsdam Institute for Climate Impact Research (known by the acronym PIK). In 1999 he received from the James S. McDonnell Foundation a $1 million Centennial Fellowship Award, one of the world's most prestigious science prizes, for his work on the role of the oceans in climate change, including the ways in which the movements and temperatures of ocean currents affect temperatures in Earth's atmosphere. His work entails gathering and interpreting vast quantities of data disseminated by other scientists, not only to determine forces at work on Earth's climate at present but also to shed light on climate changes over thousands of years in the past. He has used such information to create computer simulations to predict changes in climate in the future.

For some years Rahmstorf was a member of the Abrupt Climate Change Panel of the National Oceanic and Atmospheric Administration (NOAA), a US agency, and he currently sits on the German Advisory Council on Global Change. He was a principal author in 2007 of the Fourth Assessment Report of the Intergovernmental Panel on Climate Change (IPCC, established in 1988 by the World Meteorological Organization and the United Nations Environment Programme). The report concluded that global warming was "unequivocal" and has almost certainly resulted from human-induced emissions of so-called greenhouse gases—particularly carbon dioxide ($CO_2$). Rahmstorf has written or cowritten more than sixty scientific papers, including fourteen published in *Nature* or *Science*; twenty book chapters, in such reference works as *Encyclopedia of Ocean Sciences* (2001); and more than two dozen articles for *New Scientist* and other publications for laypeople. He has also coauthored three books: *Der Klimawandel* ("Climate Change," 2006), *Our Threatened Oceans* (2008), and *The Climate Crisis: An Introductory Guide to Climate Change* (2009). He is a cofounder and regular contributor to RealClimate, a blog written by climate scientists for both experts and laypeople that has had over two million visits in the past five years. In 2005 the blog won a Science and Technology Web Award from *Scientific American*. In an online post (October 3, 2005), the magazine's editors described RealClimate as "a refreshing antidote to the political and economic slants that commonly color and distort news coverage of topics like the greenhouse effect, air quality, natural disasters, and global warming" and "a focused, objective blog written by scientists for a brainy community that likes its climate commentary served hot."

In recent years Rahmstorf has become prominent among scientists who are vigorously challenging the views of so-called climate-change deniers, contrarians, and skeptics. He has strived to convince the public that a vast and growing body of evidence proves beyond doubt that the activities of humans have caused significant rises in oceanic and atmospheric temperatures; that failure to reverse that trend in the next few years will be catastrophic for hundreds of millions of humans and many other species of animals and plants; and that as many as a third of those species may well die out. "I wish [the skeptics and deniers] were right," he told *Current Biography*. "But unfortunately the recent data show that we have not exaggerated but underestimated the problem thus far. Arctic sea ice is vanishing much faster and sea levels are rising more rapidly than we expected just a few years ago." "I believe we will stop global warming," he said. "The question is just: how fast and at what level? I am less optimistic that we have enough political will and courage to curb emissions fast enough to limit warming to below 2 degrees Celsius, as is the goal that has now become a global consensus. . . . I won't give up hope, but I worry that we will do too little, too late."

## Early Life and Education

Stefan Rahmstorf was born on February 22, 1960, in Karlsruhe, Germany. His father, Rolf Rahmstorf, was a manager in the pharmaceutical industry, and his mother, Hildegard, was a pharmacist. At age six, his family, which includes his brother and sister, moved to the Netherlands. Rahmstorf has said that by the age of twelve he knew that he wanted to become a scientist. "I always wanted to understand how the world works," he told *Current Biography*. "I was particularly interested in astronomy and physics. I started to learn about oceanography in 1982, mid-way in my physics studies. I'd always loved the oceans and knew right away that oceanography was going to be the right field for me."

Rahmstorf attended the University of Ulm and then the University of Konstanz, both in Germany; at the latter he wrote a thesis on general relativity theory and earned a diploma in physics. During his undergraduate years he also studied physical oceanography at the University of Wales (now Bangor University) in Bangor. He next pursued a PhD in oceanography at Victoria University of Wellington, New Zealand. He conducted research while on several cruises in the South Pacific. After he earned his doctorate in 1990, Rahmstorf worked briefly as a scientist at the New Zealand Oceanographic Institute in Wellington. In 1991 he joined the Institute of Marine Sciences at the University of Kiel, in Germany. He has been working at the Potsdam Institute for Climate Impact Research since 1996 and has taught at the University of Potsdam since 2000.

## Life's Work

Most of Rahmstorf's research on the role of ocean currents in climate change concerns what is known as the thermohaline circulation (THC) of the ocean (the world's connected seas). As Rahmstorf explained in a fact sheet posted on the PIK website (2002), the THC is different from wind-driven currents and tides, which occur at or near the surface and are controlled by the forces of gravity exerted on the ocean by the moon and the sun. The THC is linked instead to differences in seawater temperature and density; denser, colder seawater sinks below seawater that is warmer and less dense. The density of seawater depends on its salinity—the concentration of salt in it. Salinity increases as seawater evaporates and sea-ice forms; it decreases with rainfall, snowfall, runoff (water from rivers and rain, snow, and irrigation that is not absorbed into land but flows into the ocean), and the addition of meltwater, from melting glaciers, ice sheets, and other ice. Thermohaline circulation—sometimes called the ocean conveyor belt or the global conveyor belt—involves the sinking of huge masses of water (in a process known as deep-water formation) in particular areas—the Mediterranean Sea, the Greenland-Norwegian Sea, the Labrador Sea, the Weddell Sea, and the Ross Sea; the spreading and upwelling of deep water; and the movements of certain near-surface currents, including the North Atlantic Drift (or North Atlantic Current). The North Atlantic Drift and the Gulf Stream (the latter of which is mostly wind-driven) bring warm water from the Gulf of Mexico up along the eastern coast of the United States and across the Atlantic Ocean. The warm water releases heat that has kept the climate of Western Europe temperate for the past 10,000 years—since the last ice age. As the water grows colder, it grows denser and sinks; that sinking drives the currents. By the time the deep, cold water has returned south, it has reached depths of one to two miles below the surface.

"The crux of the matter is that the strength of the circulation depends on small density differences, which in turn depend on a subtle balance in the North Atlantic between cooling in high latitudes and input of rain, snowfall, and river runoff," Rahmstorf explained in an article for *UNESCO Sources* (December 1997). Continued global warming, he has maintained, will upset that subtle balance. With an abnormally large input of freshwater from melting ice, for example, seawater will become less salty and thus less dense, deep-water formation in the Atlantic will slow, and the movement of the "conveyor belt" will slow and eventually stop entirely. "We can see that such breakdowns have happened before in climate history . . . ," Rahmstorf wrote for *UNESCO Sources*, although precisely what caused them is not yet known. "Pulses of freshwater entered the Atlantic, causing cold spells lasting for hundreds of years. The last cooling occurred about 11,000 years ago. . . . So the possibility of a circulation breakdown is not just in the computer, it is real." The evidence comes from painstaking examinations of cores of ice extracted from Greenland and cores of sediments extracted from the floor of the ocean. The cores, from drillings to depths of thousands of feet, show the accumulations of yearly depos-

---

**"I believe we will stop global warming. The question is just: how fast and at what level?"**

---

its, the contents of which can be analyzed to reveal climate conditions at the time that each deposit settled. (Such analyses are analogous to those of tree rings, which are added annually and reveal periods of drought, for example.)

## The Debate over Global Warming

The presence in Earth's atmosphere of carbon dioxide and other so-called greenhouse gases—methane, nitrous oxide, carbon monoxide, and (for the past 250 years or so) others produced by various industrial operations, such as sulfur hexafluoride—is not inherently harmful. On the contrary: such gases are vital for life, because—in a process dubbed the greenhouse effect in the mid-1800s—they trap long-wave radiation escaping from Earth into space, which normally balances the energy coming from the sun. In a self-regulating, natural system, the trapped radiation heats the atmosphere and keeps air temperatures at levels favorable to life as we know it. As humans began to burn fossil fuels (coal and liquid-petroleum products) to run machinery in factories and, later, to run power plants, cars, and planes, unnatural amounts of $CO_2$ and other greenhouse gases were emitted into the atmosphere. Scientists' understanding of climate and the forces that control it is incomplete (as is their understanding of weather—climatic conditions in small locales minute by minute). However, they know without any doubt that the burning of fossil fuels (and, to a much lesser extent, deforestation) has led to the increased concentration of $CO_2$ in the atmosphere. It is also known that increases in $CO_2$ concentrations lead to an increase in trapped long-wave radiation, a rise in atmospheric temperatures near Earth's surface, and higher temperatures in the surface waters of the ocean. By the early 1990s, according to Rahmstorf, global temperatures on average were 0.5 degrees Celsius higher than they were before the Industrial Revolution, and since then they have risen another 0.3 degrees Celsius.

Some of the effects of global warming are already apparent. Among the most dramatic is the loss of summer sea ice in the Arctic Ocean. "Ice extent in 2007 and 2008 was only about half of what it [was] in the 1960s," Rahmstorf wrote in his article "Climate Change—State of the Science," posted on the PIK website in 2009; "ice thickness has decreased by 20–25 [percent] just since 2001, and in 2008 the North-East Passage and North-West Passage were both open for the first time in living memory." One consequence of that melting of sea ice is that, in the past four decades, "sea levels have increased about 50 percent more than the climate models predicted," Rahmstorf told an Agence France Presse reporter (December 14, 2006). (Rahmstorf provided evidence for that conclusion in a paper, co written with six others, that was published in the May 4, 2007, issue of *Science*.) If sea levels were to continue to rise, land at or slightly above sea level would disappear under water, leaving homeless many of the millions of people who live on low-lying islands or in coastal areas. Another result of global warming is that the acidity of the oceans is increasing along with higher levels of $CO_2$. Such acidity threatens the well-being and even survival of marine ecosystems and fish and crustacean populations in many parts of the world.

The reality of anthropogenic (human-caused) global warming has been accepted as fact by the IPCC, the NOAA, the UN Environment Programme, the World Meteorological Organization, the National Academy of Sciences of the United States, the American Geophysical Union (the world's largest association of Earth scientists), the National Aeronautic and Space Administration of the United States, and dozens of other scientific government agencies as well as climatologists from around the world and the leaders of many nations. Despite that consensus, Rahmstorf complained in his paper "The Climate Sceptics"

(2005), published in conjunction with a conference sponsored by the organization Munich Re, the media continue to convey to the public the false idea that the main conclusions of the scientific community regarding global warming "are still disputed or regularly called into question by new studies." In light of the voluminous evidence produced by thousands of studies, he wrote, "it is almost inconceivable" that those conclusions "could be overturned by a few new results." The media's persistent propagation of false ideas, in his view, "is mainly due to the untiring PR [public relations] activities of a small, but vocal mixed bag of climate sceptics . . . who vehemently deny the need for climate-protection measures." Prominent among that "vocal mixed bag" are representatives of, lobbyists for, or politicians who have benefited from the contributions of corporations that sell petroleum products or other commodities whose manufacture and use leads to the emission of greenhouse gases. In "The Climate Sceptics" and elsewhere, Rahmstorf has shown that there is no scientific evidence to support any of their arguments but that there is a huge body of evidence to refute them. He has pointed out the fallacies in arguments that global temperatures are not rising; that natural processes account for nearly all the increase in $CO_2$; that $CO_2$ plays little or no role in global warming; that greater solar activity has caused increased temperatures on Earth; that the potential effects of global warming are negligible; and that the costs of measures proposed for stopping or reversing global warming by limiting $CO_2$ emissions would be prohibitively high.

The last-mentioned argument was presented in an article for *Newsweek* (August 29, 2009) by Bjorn Lomborg of Denmark, whose academic training is in political science and who teaches at the Copenhagen Business School. According to Lomborg, "Even if all industrialized nations succeeded in meeting the most drastic emissions goals, it would likely come at a huge sacrifice to prosperity." Referring to figures offered by the economist Richard Tol of the Economic and Social Research Institute in Ireland, he wrote, "Using carbon cuts to limit the increase in global temperature to 2 degrees Celsius . . . would cost 12.9 percent of [the world's gross domestic product, or GDP] by the end of the century. That's $40 trillion a year. . . . Yet, such measures would avoid only \$1.1 trillion in damage due to higher temperatures. The cure would be more painful than the illness." Lomborg suggested as far more cost-effective the use of "climate engineering": "For instance, automated boats could spray seawater into the air to make clouds whiter, and thus more reflective, augmenting a natural process. Bouncing just 1 or 2 percent of the total sunlight that strikes the Earth back into space could cancel out as much warming as that caused by doubling pre-industrial levels of greenhouse gases. Spending about \$9 billion researching and developing this technology could head off \$20 trillion of climate damage."

Rahmstorf has rejected such approaches. "First of all they do nothing to stop ocean acidification, which by itself would be enough reason to cut down our $CO_2$ emissions, unless we want to destroy our ocean ecosystems," he told *Current Biography*. "Secondly they would make the climate system inherently unstable and requiring [of] human control, which would have to be maintained for thousands of years because of the long lifetime of the $CO_2$ that is building up in the atmosphere. If you let those cooling measures slip for just a few years, the full force of $CO_2$ warming would hit us immediately. I just don't trust humans can reliably manage such a complex system for millennia— I'd much rather keep it the relatively stable self-regulating system it is now." Rahmstorf has endorsed the work of the British economist Nicholas Stern, who, after a review of other economists' studies, has estimated that the cost of cutting emissions would total 2 percent of global wealth (as measured by the gross domestic product of all nations), while the damage resulting from inaction would cost anywhere from 5 to 20 percent of global wealth, and possibly far more. In an interview posted on the Allianz Knowledge website (June 9, 2009), Rahmstorf suggested that people could "build an energy system over the next decades based primarily on renewable resources"—biofuels, solar power, and wind power—and thus avoid the use of fossil fuels, at least to a significant extent.

On November 18, 2008, in taped remarks that Rahmstorf has sometimes quoted, President-elect Barack Obama addressed the Bipartisan Governors Global Climate Summit, held in Los Angeles, California.

Obama told some thirty US governors and representatives of many nations, "Few challenges facing America—and the world—are more urgent than combating climate change. The science is beyond dispute and the facts are clear. . . . Now is the time to confront this challenge once and for all. Delay is no longer an option. . . . The stakes are too high. The consequences, too serious." The United States, which in recent years has been responsible for at least 21 percent of greenhouse-gas emissions, did not join the 185 countries that ratified the Kyoto Protocol, drawn up by thirty-seven industrialized nations and the European Union during the UN Framework Convention on Climate Change held in Japan in 1997. A legally binding treaty that went into effect in 2005, the protocol aimed to reduce greenhouse-gas emissions from 1990 levels by an average of 5 percent between 2008 and 2012. In December 2009, when the United Nations held its fifteenth conference on climate change in Copenhagen, $CO_2$ levels were continuing to increase, in part because of China's rapidly rising use of fossil fuels and the failures of nearly every other nation to cut emissions. Government heads and other delegates from 192 countries (among them Obama and Rahmstorf) attended that gathering. The agreement that resulted, known as the Copenhagen Accord, "fell short of even the modest expectations for the summit . . . ," according to the *New York Times* (December 18, 2009). "The accord drops what had been the expected goal of concluding a binding international treaty by the end of 2010, which leaves the implementation of its provisions uncertain. It is likely to undergo many months, perhaps years, of additional negotiation before it emerges in any internationally enforceable form." The *Times* also reported, "The maneuvering that characterized the final week of the talks was a sign of [global leaders'] seriousness; never before have [they] come so close to a significant agreement to reduce the greenhouse gases linked to warming the planet." The leaders agreed that global warming must never exceed 2 degrees Celsius above pre-industrial levels and that by the end of January 2010, wealthy nations must register the emissions cuts they will implement by 2020. Those nations also pledged to fund efforts by poorer nations to reduce greenhouse-gas emissions and adapt to consequences of climate change.

## Personal Life

In his free time Rahmstorf enjoys photography, yoga, and dancing. His first marriage in 1991 to the New Zealand-born actress Dulcie Smart, ended in divorce. He lives with his second wife, Stefanie, and their two children in Potsdam, where his wife owns a jewelry-design shop. They sell sterling-silver jewelry of their own design; each piece reads "OCO" or in some other way refers to $CO_2$. The Rahmstorfs use the money from each sale to buy one of the limited number of $CO_2$-emission credits issued by the European Union. They stamp each piece of jewelry with the serial number of the credit, to serve as evidence that with that purchase, the buyer has prevented the emission of one ton of $CO_2$ into the atmosphere.

## Further Reading

Kunzemann, Thilo. "Taking on Climate Change Myths and Skeptics." *Allianz Knowledge*. Allianz, 9 June 2009. Web. 8 Aug. 2012.

Sawyer, Kathy. "They're Young, Brilliant and $1 Million Richer; Foundation Rewards 10 'Geniuses of the 21st Century.'" *Washington Post* 12 Apr. 1999: A3. Print.

"Stefan Rahmstorf." *PIK-Potsdam.de*. Potsdam Institute for Climate Impact Research. Web. 8 Aug. 2012.

# Randall, Lisa

## American physicist

**Born**: June 18, 1962; New York City, New York

"The cosmos could be larger, richer and more varied than anything we imagined," the physicist Lisa Randall wrote in her book, *Warped Passages: Unraveling the Mysteries of the Universe's Hidden Dimensions*. In an article written for the *London Daily Telegraph* (June 1, 2005), Randall explained, "My recent studies of extra dimensions of space, beyond the familiar 'up-down,' 'left-right' and 'forward-backward,' have made me more than usually convinced that they must really exist." "I really don't see any reason why these extra dimensions should not exist," she told Michael Brooks for *New Scientist* (June 18, 2005). "At the moment we see things over a very limited range of distance and energy. Every time we have looked beyond the boundaries of what we could see before, we have found new phenomena." Randall was a professor of physics at the Massachusetts Institute of Technology (MIT) when, in 1998, in collaboration with the particle physicist Raman Sundrum, who is now at Johns Hopkins University, she developed a theory to explain why gravity is by far the weakest of the four fundamental forces of nature. (The others are the electromagnetic force, which acts on atomic particles carrying an electrical charge; the strong nuclear force, which holds atomic nuclei together; and the weak nuclear force, which is associated with particle emission and nuclear decay.) Gravity, Randall argued, seems weaker than the other forces because only a small portion of it is being exerted in the three-dimensional universe with which humans are familiar (in which time is considered a fourth, nonspatial dimension); the rest, she hypothesized, might be found in other dimensions—dimensions that, as she wrote for the *Daily Telegraph*, "might be tiny, far smaller than an atom, or they might be big, or even infinite in size." Furthermore, she theorized, those dimensions might exist beyond the boundaries of our universe, in other universes; that is, our universe may simply be what has been dubbed a brane and may be surrounded by other branes. Two 1999 papers in which Randall and Sundrum presented their unprecedented ideas are among the most frequently cited in the professional literature of physics, and many scientists who at first expressed skepticism about their theories have come to embrace them. Since 2001 Randall has expanded on her ideas along with the University of Washington particle physicist Andreas Karch, among others. In the widely acclaimed *Warped Passages*, Randall wrote, "Research into extra dimensions has also led to other remarkable concepts—ones that might fulfill a science fiction aficionado's fantasy—such as parallel universes, warped geometry, and three dimensional sinkholes. I'm afraid such ideas might sound more like the province of novelists and lunatics than the focus of real scientific inquiry. But outlandish as they might seem at the moment, they are genuine scientific scenarios that could arise in an extra-dimensional world." "I want to be clear that this is science and not just science fiction," Randall said to Jane Ganahl for the *San Francisco Chronicle* (September 1, 2005). "That even though these are still theoretical ideas, they are consequences of Einstein's theory of gravity." In a review of *Warped Passages* for the *Times Higher Education Supplement* (June 3, 2005), Lorna Kerry wrote, "Anyone accusing [Randall] of straying into fantasy would be brought up sharp by the detailed grounding in physics she provides. . . . Reading *Warped Passages* gives an insight into how people must have felt on discovering that the Earth was not flat." Randall, who has taught at Harvard University since 1999, hopes that an experiment slated to be conducted at the European Organization for Nuclear Research (CERN) will offer evidence for the existence of additional dimensions.

Science, Randall told Sarah Baxter for the *London Sunday Times* (June 19, 2005), focuses as much on questions as it does answers, which is what "makes it fun." "Very often research is about finding the small glitch or inconsistency that is at the root of the really big issues," she said. According to Raman Sundrum, Randall has "a great nose" for identifying fruitful areas of investigation. He said to Dennis Overbye for the *New York Times* (November 1, 2005), "It's a mystery to those of us—hard to understand, almost to the point of amusement—

how she does it without any clear sign of what led her to that path. She gives no sign of why she thinks what she thinks." Similarly, Andrew Strominger, a professor of physics at Harvard, told Adrian J. Smith for the *Harvard Crimson* (January 6, 2006), "She is very intuitive. She often understands things in her own mind before she is able to formulate it mathematically." Randall told Michael Brooks that one reason she devoted so much time away from her research to write *Warped Passages* was that she "wanted to show why physicists think about things like extra dimensions, and how they might tie into 'real' observable phenomena." Another reason, she told Anna Fazackerley for the *Times Higher Education Supplement* (June 3, 2005), was that she "thought it was important . . . to get the message across that you don't have to look like a weirdo to do science."

## Early Life and Education

The middle of the three daughters of a salesman for an engineering company and a schoolteacher, Lisa Randall was born on June 18, 1962, in the New York City borough of Queens. Her younger sister, Dana Randall, teaches math and computer science at the Georgia Institute of Technology. By her own account, Lisa Randall was drawn to mathematics at an early age; as she told Sarah Baxter, "I was looking for certainty. What I liked [about math] was that there were definite, nice, neat answers." Randall attended Stuyvesant High School, a New York City public school for academically gifted students who are admitted on the basis of their performance on a competitive examination. While enrolled in her first physics class at Stuyvesant, as she recalled to an interviewer for *American Scientist* (September 28, 2005), she "began to think science could provide a better outlet for my interests, one that is more grounded in the physical world." Randall was the first female student to serve as the captain of Stuyvesant's math team. As a senior she tied for first place in the 1980 national Westinghouse Science Talent Search (now named for a different sponsor, Intel), with a project on complex numbers; the oldest and most prestigious precollege science competition in the United States, recognition in the Science Talent Search is considered by many to be equivalent to a "junior Nobel Prize."

After her high-school graduation, Randall entered Harvard University in Cambridge, Massachusetts. During her summer breaks she conducted research, in 1981, at the Smithsonian Astrophysical Observatory, which is affiliated with Harvard; in 1982, at the IBM facility at Poughkeepsie, New York, and then the Fermi National Accelerator Laboratory (FNAL), near Chicago, Illinois; and in 1983, at Bell Labs in New Jersey. She earned a BA degree in physics from Harvard in 1983 after only three years. She remained at Harvard to pursue a PhD in particle physics, a subject that had captured her imagination during her stint at FNAL. Her PhD adviser was Howard Georgi, a strong supporter of women in science; her dissertation, entitled "Enhancing the Standard Model," about the theory of fundamental particles and how they interact, was based on collaborative work of Georgi and Sheldon Glashow, the latter of whom won the Nobel Prize in Physics in 1979. She was awarded a PhD in 1987, by which time she had coauthored seven papers published in physics journals. Next, as a postdoctoral fellow, she worked at the University of California at Berkeley (1987–1989), the Lawrence Berkeley Laboratory (1989–1990), and Harvard (1990–1991).

### Life's Work

In 1991 Randall joined the faculty of MIT in Cambridge as an assistant professor of physics; four years later she was promoted to associate professor. Earlier, in 1992, she had won an Outstanding Junior Investigator Award from the Department of Energy; a Young Investigator Award from the National Science Foundation; and an Alfred P. Sloan Foundation Research Fellowship. While at MIT she wrote or co wrote some fifty papers. She left that school in 1998 to teach at Princeton University in New Jersey, where she became the first female professor of physics to earn tenure. In 2000 she returned to MIT; there, she held the distinction of being the school's first female theorist in particle physics.

Meanwhile, in 1998, Randall and Raman Sundrum—who was then a postdoctoral fellow at Boston University and whom Randall had worked with earlier—had begun to meet at an ice cream parlor at the MIT student center to discuss some baffling problems regarding gravity. Sundrum, who is now a professor at Johns Hopkins University, "had already thought about branes and extra dimensions, and he was an obvious person to join forces with," Randall explained to Marguerite Holloway for *Scientific American* (October 2005). Describing the questions that she and Sundrum grappled with, Randall told John Crace for the *London Guardian* (June 21, 2005), "The Standard Model of particle physics describes forces and particles very well, but when you throw gravity into the equation, it all falls apart. You have to fudge

---

**"The cosmos could be larger, richer and more varied than anything we imagined."**

---

the figures to make it work. We know how gravity works, but no one had properly answered the question of what determines its strength. Why is it so much weaker than standard theory would predict?" Gravity is by far the weakest of the four fundamental forces of nature—trillions of trillions of trillions of times weaker. "Gravity might not appear to be all that weak when you're hiking up a mountain, but bear in mind that the gravitational force of the entire Earth is acting on you," Randall wrote for the *Daily Telegraph*. "Think how feeble gravity must be for you to counter the force of the much larger Earth when you pick up a ball." In another common example of gravity's weakness, a magnet can hold a paper clip many times its size above the ground, even though Earth's gravity is pulling at it.

Randall and Sundrum theorized that gravity may be so weak because much of its force is being exerted in dimensions that humans are unable to perceive. While humans have traditionally thought of themselves as functioning spatially in three dimensions, Randall views dimensions as "the number of quantities you need to know to completely pin down a point in a space," as she put it in *Warped Passages*. Randall told Ira Flatow for the National Public Radio program *Talk of the Nation* (September 30, 2005), "One of the things that makes [the idea of additional dimensions] difficult to get across is that you can't picture it. We are not physiologically designed to picture more than three dimensions. It doesn't mean they're not there, but we certainly can't just picture them very simply. What we can do is we can try to extrapolate our ideas mathematically or in words, and we can find shortcuts for trying to picture things, the same way we try to picture three-dimensional worlds on two-dimensional pieces of paper." The idea that the universe is composed of more than three spatial dimensions is a crucial element of string theory, a central component of modern physics. String theory was developed as a way of reconciling classic, Newtonian physics, which explains the basic laws of motion, and quantum mechanics, which is concerned with the movement of atomic particles. According to Randall in the *Daily Telegraph*, string theory "doesn't naturally describe a world with three dimensions of space. It more naturally suggests a world with many more, perhaps nine or ten. A string theorist doesn't ask whether extra dimensions exist; instead, two critical questions that a string theorist asks are: where are they and why haven't we seen them?" String theorists had so far posited that the extra dimensions are undetectable by humans, theoretically as well as in actuality, because each is "rolled up" into a tiny bit of space far smaller than an atom.

"Many physicists are willing to overlook the lack of experimental evidence [for the validity of string theory] because they believe that string theory will eventually reconcile quantum mechanics, which governs atoms and all other particles, with general relativity, which describes how matter and gravity interact on the very largest scales," the science writer and editor Tim Folger noted in a review of *Warped Passages* for the *New York Times* (October 23, 2005). "Randall, though, argues that without any experimental feedback, string theorists may never reach their goal. She prefers a different strategy, called model building. Rather than seeking to create an all-encompassing theory, she develops models—mini-theories that target specific testable problems and that might then point the way to a more general theory." Adopting that approach, in 1999 Randall and Sundrum published a pair of papers, "Large Mass Hierarchy from a Small Extra Dimension" and "An Alternative to Compactification," in the journal *Physical Review Letters*, the first of which discusses the weakness of gravity in the known universe (the hierarchy problem, as it is labeled), and the second of which proposes that the existence of so-called branes would account for the existence of an infinitely large extra dimension. A brane (the word, coined by cosmologists, comes from "membrane," something that, on the cellular level, acts as a barrier) is an "object in higher-dimensional space that can carry energy and confine particles and forces," according to Randall's book. Human beings exist on a "weak brane," while the brunt of gravity's force would be exerted on a stronger brane. "The world of branes is an exciting new landscape that has revolutionized our understanding of gravity, particle physics, and cosmology," Randall wrote in *Warped Passages*. "Branes might really exist in the cosmos, and there is no good reason that we couldn't be living on one." Randall told Gary Taubes for the website Special Topics that her work with Sundrum "went in the face of what everyone who has studied gravity has believed. We always thought if we have extra dimensions, we had to do what's called compactifying them, which is to say they curl up so that they can't be seen from our point of view. . . . It wasn't thought to be possible to have a theory with extra dimensions that wouldn't compactify and still have the physics reduce to four-dimensional physics here. So our theory, that you could have an infinite extra dimension that wasn't compactified, was a radical departure from conventional wisdom." As Anna Fazackerley wrote for the *Times Higher Education Supplement* (June 3, 2005), Randall's work with Sundrum "transformed the way that theoretical physicists think about the underlying structure of space." Randall wrote in *Warped Passages*, "Physicists have now returned to the idea that the three-dimensional world that surrounds us could be a three-dimensional slice of a higher-dimensional world."

**Kaluza-Klein Particles and Extra-Dimensional Theory**

Fundamental to Randall's theory about extra dimensions is the presumed existence of what are called Kaluza-Klein particles. As Randall wrote in *Warped Passages*, "KK [Kaluza-Klein] particles are the additional ingredients of an extra-dimensional universe. They are the four-dimensional imprint of the higher-dimensional world. . . . Just as Flatlanders [characters from Edwin A. Abbott's 1884 book *Flatland*], who see only two spatial dimensions, could observe only two-dimensional disks when a three-dimensional sphere passed through their world, we can see only particles that look like they travel in three spatial dimensions, even if those particles originated in higher-dimensional space. These new particles that originate in extra dimensions, but appear to us as extra particles in our four-dimensional space-time, are Kaluza-Klein (KK) particles." According to the Fermi National Accelerator Laboratory website (www-cdf.fnal.gov) that discusses the search for extra dimensions by observing what is called missing energy, a Kaluza-Klein particle is a graviton—"the particle that mediates the force of gravity." If two atomic particles on the brane produce sufficiently high energy when they collide—specifically, in a collision in a high-energy particle accelerator—they will produce a graviton. "The graviton flies out of the brane . . . carrying away energy and momentum. . . . An observer on the brane witnessing the outcome of the collision would see the usual particles produced in such experiments, with a large imbalance of energy and momentum. Instead of detecting the Kaluza-Klein graviton directly, we observe its 'missing energy' signature." "The most exciting feature of any extra-dimensional theory that explains the weakness of gravity is that if it is correct, we will soon find out," Randall wrote in *Warped Passages*. An experiment designed to test the theory will be carried out at CERN, on the border of France and Switzerland near Geneva, where the most powerful atom-smasher on Earth—what has been dubbed the Large Hadron Collider—is located. If the experiment turns out as Randall hopes it will, the collision of high-energy particles will produce evidence of additional dimensions. "If this is the way gravity works in high-energy physics, we'll know about it," Randall told Dennis Overbye.

Randall completed writing *Warped Passages* (2005) in three years. The 512-page book, whose title refers to the warping of the so-called space-time continuum, offers a history of modern theoretical physics along with a description of her own theories. Among the many enthusiastic reviews it earned was that of Brian Cox, who wrote for the *Times Higher Education Supplement* (September 16, 2005), "It is difficult stuff, but [Randall] makes it as simple as possible without diluting the content." In an assessment for *Library Journal* (September 1, 2005), Sara Rutter wrote, "To explain and illustrate the complex models and mathematical calculations used to develop groundbreaking new theories in physics, Randall employs stories, analogies, and drawings. In this way, she is like an extraordinarily smart and lively college professor working to engage her students in the excitement of discovery."

In 2005 Lawrence Summers, then the president of Harvard University, appointed Randall to a school task force whose mission was to recruit female science professors. Shortly before that happened Summers had ignited a fierce controversy on and off the Harvard campus by remarking in a speech that the significantly greater representation of men in the hard sciences might be attributed to inborn differences in ability between men and women. Randall disputed that idea, telling Michael Brooks, "There are lots of social factors that influence performance so it's impossible at this point to isolate innate differences in intelligence. You can only say something reliable about innate differences in scientific ability if the differences are so big they cannot be explained by social factors. And so far as we can tell, they are not." She told Matin Durrani for *Physics World* (May 2005, online) that women "just need a level playing field. There really are prejudices [against women], both inadvertent and deliberate, and if we get rid of those, we don't need any targets"—that is, minimum numbers of women on the faculty of any department of physics.

Among her many awards and honors, Randall has won the Klopsteg Award (2006), the American Physical Society's Julius Edgar Lilienfeld Prize (2007), the Radcliffe Fellowship Award (2009), the Erna Hamburger Prize (2010), and the Andrew Gemant Award from the American Institute of Physics (2012).

Many reporters have used words such as "glamorous" to describe Randall's looks. In 2007, *Time Magazine* included Randall in its "Time 100" list of the most influential people in the world. In her leisure time she enjoys skiing, rock climbing, and watching movies.

### Further Reading

Brooks, Michael. "Interview: The Final Frontier." *New Scientist* 18 June 2005: 88. Print.

Crace, John. "Lisa Randall: Warped View of the Universe." *The Guardian* 21 June 2005: 20. Print.

Dizikes, Peter. "Across the Universe." *Boston Globe* 4 Sept. 2005: E1. Print.

# Smith, Amy

## American mechanical engineer

**Born**: November 3, 1962; Lexington, Massachusetts

"Problem solving has always been in my blood," the mechanical engineer and inventor Amy Smith told Elizabeth Karagianis for the Massachusetts Institute of Technology (MIT) publication *Spectrum* (Spring 2000). "I'm the kind of person who will walk into a restroom, see a broken sink and fix it instead of complaining that someone else should do it." Far from confining herself to improvements of plumbing or anything else in common use in the United States, Smith has devoted her extraordinary ingenuity and energy not to advancing technologies but, as Kari Lynn Dean wrote for *Wired News* (October 11, 2004), to "using old technology in fresh ways," with the goal of improving the lives of thousands of poor people in the most impoverished places on Earth. Indeed, thousands of people have already benefited from her three most widely known inventions: a relatively inexpensive, easy-to-repair portable mill for grinding grain into flour; a nonelectrical device called a phase-change incubator, for identifying potentially harmful microorganisms in drinking water; and an oil-drum "kiln" for converting waste sugarcane stalks (stalks previously pressed for their juice) into charcoal—a fuel that is vitally needed in parts of the world where people still cook over wood fires but where supplies of firewood are rapidly dwindling. "Looking at things from a more basic level, you can come up with a more direct solution, and a lot of people go well, duh, that's really obvious!" Smith said to Dean. "But that's what you want: people saying it should have been done that way all along. It may sound small in theory, but in practice, it can change entire economies." She also told Dean, "A lot of people look at where technology is right now and start from there. . . . If you go back to the most basic principles, you can eliminate complexity. The stuff I do is just very simple solutions to things, which is critical when you are developing applications for the third world." A former Peace Corps volunteer who now teaches at MIT, Smith has won several prestigious awards for her work, most notable among them the remunerative MacArthur Fellowship. In an interview with a reporter for *PR Newswire* (February 9, 2000), she observed, "Necessity is the mother of invention, but it has often struck me that the most needy are often the least empowered to invent." Referring to her MacArthur grant, she said to Dean, "I've always wanted to have funding to help the people with good ideas who I see and think, 'If only they had the resources, they could do this cool project.' I'd like to use a significant portion of the money to enable people who have the potential to make a significant impact in places like Haiti and India, the ability to just say, 'Hey, that sounds good, let's do it.'"

## Early Life and Education

The second of four children, Amy Smith was born on November 3, 1962, in Lexington, Massachusetts, where she grew up. Her mother was a junior-high-school math teacher; Smith has said that her mother always taught her, as Elizabeth Karagianis paraphrased Smith's words, "If you see a problem, do something about it." Smith's father, Arthur C. Smith, is a professor of electrical engineering at MIT; for five years he served there as the dean for undergraduate education and student affairs, and for two years, he was chairman of the faculty. Smith's older sister, Abby Smith, teaches marine science at the University of Otago in New Zealand. Smith also has a younger sister and a younger brother. In interviews she has recalled that her family had stimulating conversations at mealtimes, on such subjects as ways to prove the Pythagorean theorem (a proposition fundamental to Euclidean geometry). During the year that she would have spent as a second-grader, Smith attended a village school for boys in northwestern India, where her family lived while her father taught at an Indian technical institute. (Classes in the boys' school, but not the girls', were taught in English.) That experience "set a lot of things in motion for her," as her father remarked to a reporter for the *Boston Business Journal* (February 25, 2000). "It's very different from growing up in a Boston suburb." Her father described Smith as having "always had a different view of what was important in the world." For seven years after she became old enough to babysit, Smith contributed half her earnings to UNICEF.

At Lexington High School Smith devoted much time to musical pursuits as a member of the band, the orchestra, the chorus, the consort choir, and the madrigals group. She also participated in many activities at the Unitarian-Universalist Follen Community Church in Lexington. After she completed high school in 1980, she enrolled at MIT. As an undergraduate she played on the campus basketball, water polo, and volleyball teams. She earned a BS degree in mechanical engineering in 1984. During the next two years, she held a part-time engineering-design job; she also volunteered at a food bank and a soup kitchen, coached Special Olympics hopefuls, and tutored students in inner-city high schools.

In 1986 Smith joined the Peace Corps for a four-year stint—twice the normal commitment. She was assigned to work in Botswana, a land-locked nation in south-central Africa. For two years she served as a teacher of math, science, and English at Itekeng Community Junior Secondary School in the city of Ghanzi. "It's clear that the algebra I taught kids there isn't going to change the whole world," she told Elizabeth Karagianis. "But the fact that the students knew I cared about them made it possible for many of them to continue their educations." For the next two years, as a Ministry of Agriculture regional beekeeping officer, Smith trained farmers in raising bees for honey. In 1988 she was named the JFK Peace Corps Volunteer of the Year in Africa (an honor named for John F. Kennedy, who as president launched the corps). "In the Peace Corps, the people are your social life," Smith told Karagianis. "You don't go to movies. You don't go to concerts. You go to people's houses and you talk to them. The big thing I learned is that people are the most important thing in the world." That sentiment was reinforced when, while Smith was overseas, her mother died. Her death, Smith said, "made me realize you must do what you can to help people. You never know when it will be too late." When she returned briefly to the United States to attend her mother's funeral, she was thrust from one of the world's poorest countries to an affluent American community in the United States, where entire aisles of the local supermarket offered variations on a single product. The jolt she felt at the contrast contributed to her later decision to continue her education and use her engineering skills for the benefit of those living in impoverished parts of the world.

In the early 1990s Smith enrolled in the graduate program in mechanical engineering at MIT. While in Botswana she had occasionally ground grain into flour, using her own muscle power to pound it, as countless women in developing nations have done for millen-

> **"Necessity is the mother of invention, but it has often struck me that the most needy are often the least empowered to invent."**

nia. Producing flour in that way "is one of the most labor-intensive tasks performed by women in rural areas of developing countries," as Smith noted on her MIT website; the work often consumes four hours or more each day. In a small fraction of the time and with far less effort, the same job can be accomplished with a commercially produced, motorized appliance called the hammermill. A hammermill is relatively costly, however, and the steel screen that separates the flour from unground particles is its most fragile part; replacements for broken screens "cannot be produced locally," as Smith explained, and they, too, are expensive. As an alternative (and as her master's-degree project) Smith devised a mill that, in her words, "take[s] advantage of the differences in the aerodynamic properties of the particles" to separate the flour from larger particles; there is no screen, and any part that breaks can be repaired locally. The cost of producing the mill is approximately a quarter of that of manufacturing the traditional hammermill, and its operation requires only 30 percent as much energy. Moreover, as she observed when she tested it in a village in Senegal in West Africa, her mill produces flour ten times faster than manual grinding, and the flour it produces has been deemed superior to what emerges from a hammermill. Smith earned an MS degree in mechanical engineering from MIT in 1995. As a student in MIT's Technology and Policy Program, she received a second master's degree, in 2001.

### The Phase-Change Incubator

Earlier, Smith had also invented what is known as the phase-change incubator. The idea for it came to her in 1997, after a visit to Uganda, where she had gone specifically to find "appropriate design problems" for an undergraduate course she was teaching, as Denise Brehm reported in an MIT press release (November 24, 1999). In Uganda, Smith had met a volunteer working for the African Community Technical Service who was testing for the presence of bacteria in local water supplies. In large parts of Uganda, as in vast areas of other parts of the developing world, virtually all available drinking water is contaminated with bacteria. In order to destroy them with the proper chemicals, the type or types of bacteria must be identified, by placing a sample of the water in question on a growth medium (a substance that stimulates bacteria to multiply rapidly), which must remain at a specific elevated temperature for at least twenty-four hours. For decades, testers had relied on equipment powered by electricity (from a generator or batteries) to raise the temperature and keep it at the proper level. But sources of electricity do not exist in many of the places that lack clean drinking water.

To fabricate a new type of incubator, Smith looked for a substance that could be heated over a wood or coal fire to a specific temperature and then, with proper insulation, retain the heat for the requisite twenty-four hours. She found it by trolling an 855-page chemical-supply catalog for materials that were neither poisonous nor expensive and that changed from the liquid phase to the solid phase at that particular temperature (about 111 degrees Fahrenheit, or 44 degrees Celsius)—the "phase-change" temperature. Just as the temperature of water that is turning to ice remains at 32 degrees Fahrenheit (0 degrees Celsius) during the transition, the wax-like substance that Smith selected for the incubator remains at about 111 degrees Fahrenheit during its transformation from liquid back to solid—a process that takes about twenty-four hours. The components of Smith's incubator, which is about one cubic foot in volume, are a rigid container thickly lined with polyurethane insulation, into which fits an aluminum cylinder filled with the wax-like substance and capped by an aluminum holder for test tubes or petri dishes. The water sample to be tested is placed on a growing medium in the test tube or petri dish. To test it for bacteria, the user first heats the filled cylinder over a fire. When the wax-like substance melts, the cylinder, with the holder at its top, is placed inside the box. After twenty-four hours, all bacteria present will have reproduced sufficiently to form visible colonies, thus enabling the user to identify their species. According to Brehm, "A liquid-crystal top that changes color with the temperature fluctuation will probably be used to alert workers that the chemical compound has reached the appropriate temperature and/or cooled again to room temperature." In a television interview with Beverly Schuch and Lauren Thierry for the CNNfn program *Business Unusual* (May 25, 2000), Smith mentioned other possible applications for the phase-change incubator, such as transporting vaccines and incubating blood samples or tissue cultures.

In 1999 Smith's phase-change incubator won the B. F. Goodrich Collegiate Inventors Award, which carried a stipend of $20,000. The next year Smith won the Lemelson-MIT Student Prize for Inventiveness for her development of the grain mill and the incubator. As quoted by the Associated Press and *Local Wire* (February 9, 2000), Lester Thurow, chairman of the Lemelson-MIT Awards Program, said upon announcing Smith's receipt of the honor, "While technology is often seen as increasing the 'digital divide,' technology is also needed to decrease that divide. Amy Smith is the perfect example of an inventor-innovator who's using technology to close that gap." Smith planned to invest the $30,000 prize money in marketing her inventions, first to small community-development organizations and later to such major relief agencies as the Red Cross and the World Health Organization.

Also in 2000 Smith joined the faculty of MIT as an instructor at the university's Edgerton Center in Cambridge, Massachusetts. In collaboration with the MIT Public Service Center, she created Public Service Design Seminars, in which students identify technological problems in developing countries and then work both on site and in MIT design laboratories to come up with solutions. She also founded the MIT IDEAS Competition (the acronym stands for "Innovation, Development, Enterprise, Action, and Service"), which, according to an MIT website, "encourages teams to develop and implement projects that make a positive change in the world." Judges evaluate entries as to their originality, feasibility, sustainability, and value to the community for which they have been designed. Winning entries in recent years include a computerized system for assisting blind or visually impaired pedestrians; a community-based telephone service that helps non-English-speaking immigrants; and a more-effective system for anticipating floods and warning those in areas that may be affected. Relevant travel expenses of competitors is paid or partially covered by MIT Public Service Fellowships or through another avenue at MIT.

In September 2004 Smith learned that she had won a MacArthur Fellowship, better known as a "genius grant," not only for what she had already invented but also in recognition of her potential for continuing to help developing areas technologically. The fellowship carries a stipend of $500,000, to be delivered in five yearly installments with no strings attached; that is, Smith has the freedom to use it in any way she wants, and she does not have to reveal how it is spent. She told various interviewers that she intended to use much of the money to supplement the funding she was already receiving for her work.

In another project, Smith and her coworkers found a way to create charcoal from agricultural waste for use as a fuel in Haiti, the poorest country in the Western Hemisphere. Hundreds of years ago Haiti's lush forests provided a seemingly boundless supply of wood for fuel. Beginning in the 1600s French settlers denuded huge tracts to create plantations, mostly for growing sugar cane, with African slaves providing the manpower. During the twentieth century Haiti's human population rose exponentially, creating an ever-growing demand for wood and a jump in logging operations. During hurricanes, the deforestation and the resulting soil erosion led to far more severe flooding and far greater destruction of remaining trees than in past centuries. Currently, forests remain on only 2 to 3 percent of the land. Reforestation efforts have so far proved futile in Haiti because of people's never-ending need for fuel. Thus, there was a pressing need for an alternative fuel. Smith and her students devised a way of producing charcoal from stalks of sugarcane, which are discarded after all the juice has been squeezed out. As described in "Fuel from the Fields: A Guide to Converting Agricultural Waste into Charcoal Briquettes Fuel," coauthored by Smith and posted on an Edgerton Center website, the materials and equipment needed are unwanted canes (called bagasse), which have been completely dried and, preferably, chopped up; an empty oil drum, slightly modified with a chisel or other tool; a few bricks; cassava flour; water; and the makings of a small fire. Stripped to its essentials, the process entails briefly heating the bagasse and then depriving it of oxygen in a completely sealed oil drum, where after twenty-four hours or so it turns into pieces of charcoal, each no bigger than about one-tenth of an inch on a side. After the charcoal is cooled, it is mixed with a paste consisting of cassava flour and water and then formed into briquettes, either by hand or with a tool. A major benefit of the charcoal, in addition to the ease and minimal cost of its manufacture, is that, unlike materials for traditional wood fires, it does not produce smoke when burned. In Haiti and many other places around the world, the smoke from indoor wood-burning cooking fires often causes respiratory diseases in babies and children; indeed, untold thousands of children die from such diseases every year.

In 2010, *Time Magazine* included Smith in its annual "Time 100," a list of the "most influential people in the world." Smith has also received the 2008 Breakthrough Award from *Popular Mechanics* magazine and the 2011 Olympus Innovation Award for her innovative teaching methods and projects.

While not working in the field, Smith devotes much of her time to her church; she has taught Sunday school, led programs that provide food for the hungry in Boston, and worked in a high-school exchange program on Hopi and Navajo reservations. In her leisure time she enjoys playing musical instruments, including the penny whistle, guitar, tenor saxophone, and Navajo flute.

## Further Reading

Daley, Beth. "'Genius' Has Its $500,000 MacArthur Awards Three Hub Residents Among 23 Recipients." *Boston Globe* 28 Sept. 2004: B1

Karagianis, Liz. "Inventing Solutions." *Spectrum*. Massachusetts Institute of Technology, Fall 2000. Web. 8 Aug. 2012.

Kennedy, Pagan. "Necessity Is the Mother of Invention." *New York Times Magazine* 30 Nov. 2003: 86. Print.

"Lemelson-MIT Awards Two $30,000 Student Prizes for Innovation." *PR Newswire* 9 Feb. 2000.

# Soderberg, Alicia M.

## American astrophysicist

**Born**: August 18, 1977; Boston, Massachusetts

"I truly won the astronomy lottery," the astrophysicist Alicia M. Soderberg said of the moment on January 9, 2008, when a star in a far-off galaxy exploded "right in front of my eyes." Soderberg, who said those words during an interview with Dennis Overbye for the *New York Times* (May 22, 2008), added, "We caught the whole thing on tape, so to speak." With an estimated 70,000 million million million (70 sextillion) stars in the observable universe—more than the number of grains of sand on our planet—the explosion of a star, known as a supernova, is not a rare event. But before January 9, 2008, it was unprecedented for anyone to witness such an explosion at the very moment when it could first be seen on Earth—when the light it created millions of light-years earlier reached telescopes on Earth; previously, astronomers had first noticed a new supernova days or weeks after the initial blast might have been visible. The supernova Soderberg observed was SN 2008D, in the constellation Lynx in the galaxy NGC 2770, whose distance from Earth is about 100 million light-years (one light-year is about 5.9 trillion miles). That Soderberg was the person who alerted other astronomers of the birth of that supernova was only partly due to luck; her coup can also be attributed to her expertise regarding supernovae and related phenomena known as X-ray flashes and gamma-ray bursts and, perhaps even more, to her unstinting devotion to unraveling the mysteries of the universe. During her undergraduate years at Bates College, she secured funds from the National Science Foundation and the Howard Hughes Foundation and a scholarship from the Los Alamos National Laboratory to augment her studies and conduct research at institutions and observatories in Chile, Hawaii, and Puerto Rico, among other places. By the time she graduated from college in 2000, Soderberg had identified nine new supernovae. When she detected SN 2008D, Soderberg was affiliated with both the Department of Astrophysical Sciences at Princeton University in New Jersey, and the Carnegie Observatories of the Carnegie Institution for Science in Pasadena, California. She has since joined the Harvard-Smithsonian Center for Astrophysics in Cambridge, Massachusetts.

Astronomers have learned that the life spans of stars and the nature of their deaths depend upon their masses. About 99 percent of stars in the observable universe—that is, as much of the universe as we can detect with our most powerful instruments—are in the first size category, about eight times the mass of the sun or smaller. In the process of fusion that occurs in the cores of those and all other stars, hydrogen nuclei fuse into helium nuclei; after billions of years the helium nuclei fuse into carbon nuclei. The fusion process stops at that point in stars in the first size category: the stars die and become white dwarfs—lumps of carbon that are about the size of Earth (but far more dense); as the cores cool the stars get increasingly dim. (Theoretically, their luminosity will eventually end, and they will become black dwarfs. No black dwarf has been detected, because the universe has not existed long enough for a white dwarf to turn into a black dwarf.) Stars in the next size category are more than eight times but less than forty times the mass of the sun. In such stars, not only the hydrogen and helium nuclei fuse; the carbon nuclei fuse, too, to become iron. That fusion process creates an insurmountable increase in the pull of gravity from the star's core, and it ends with the collapse of the core and then a tremendous shockwave that in turn sparks the enormous explosion that marks the start of the death of the star. The energy emitted in the explosion is equivalent to that of trillions of nuclear bombs set off simultaneously; it produces a blast of luminous X-rays and, an hour or two later, stupendously bright light rays, as pieces of the star shoot forth at speeds as great as 5,000 miles per second. When such stars die they generally become neutron stars. Stars in the third size category are at least forty times the mass of the sun; their deaths generally result in black holes.

SN 2008D was a star in the second category. The blast of X-rays from SN 2008D was recorded by a satellite, sent into orbit by the National Aeronautics and Space Administration (NASA), that detects X-rays, gamma rays, and ultraviolet light in space. At Soderberg's request, in early January 2008 that satellite, called *Swift*, was aimed at a month-old supernova in NGC 2770 called SN 2007uy. The information about an X-ray blast from another part of that galaxy—which Soderberg, who was giving a lecture at that time, received in the form of an email message—electrified her, partly because it is extremely unusual for one supernova to occur virtually on the heels of another in the same galaxy; on average, supernovae occur once every one hundred years in any particular galaxy. The X-ray blast "flooded the satellite's instrument," Seth Borenstein wrote for the Associated Press (May 22, 2008), its exceptional brightness being similar to that of an image in a digital camera pointed at the sun. Soderberg immediately contacted many of her colleagues worldwide in the field of high-energy astrophysics, who in turn arranged to have both ground-based telescopes and satellites pointed at the site of the blast. Within two hours the supernova became visible as a brilliant light, some 100 billion times brighter than the sun, emitted from radioactive elements released during the explosion. Scientists continued to record data from the site continually for the next month, acquiring much new information about the death throes of a star. One reason why supernovae fascinate astronomers is that most of the chemical elements that are the building blocks of new stars and planets—and all the living organisms that inhabit Earth—are created by supernovae.

The paper in which SN 2008D was first described in the scientific literature, published in *Nature* (May 22, 2008) and titled "An Extremely Luminous X-ray Outburst at the Birth of a Supernova," lists as coauthors the names of forty-two astronomers in addition to Soderberg. They include Edo Berger, a postdoctoral research associate at Princeton who was working with her at the time, as well as scientists from England, the Netherlands, Germany, Canada, South Africa, China, Taiwan, Israel, and ten different US states. "It pays to be alert and to follow things up with energy," Robert Kirshner, a supernova specialist at Harvard University and one-time mentor of Soderberg's, observed, as quoted in a Princeton University news release (May 21, 2008). "Alicia has always been an energetic phenomenon: She has a keen sense of the possibility of finding something new by moving fast to get the data, and SN 2008D is just one more case where this has paid off." Soderberg's honors include the 2009 Annie Jump Cannon Award from the American Astronomical Society, which recognizes "outstanding research and promise for future research."

## Early Life and Education

The only child of Nancy and Jon Soderberg, Alicia Margarita Soderberg was born on August 18, 1977, in Boston, Massachusetts. Raised in Falmouth, on Cape Cod, Alicia Soderberg was encouraged from an early age to explore the world of science. She told *Current*

---

**"You have to be prepared twenty-four hours a day, seven days a week for the unexpected discovery of your career. My ultimate accomplishment would be to disprove the currently accepted notion of 'recipe for success.' A bit of hard work and quick thinking can actually be your winning ticket."**

---

*Biography*, "My parents always emphasized science and academics in general. There was never an emphasis on social status, just a constant push to succeed academically. This was true not just of my parents but also of Falmouth in general. . . . There were always very bright professors teaching our physical science classes and lots of opportunity to do research in the summer months." As a high-school student, Soderberg spent her summers conducting research on water pollution affecting Cape Cod ponds.

After her graduation from Falmouth High School in 1995, Soderberg enrolled at Bates College in Lewiston, Maine, with the assumption that she would aim for a career in environmental science. Instead, she fell in love with astronomy while taking a basic course in that subject and pursued a double major in astronomy and physics. Because Bates College offered few additional astronomy classes, Soderberg spent three academic terms at other institutions, including the Harvard Smithsonian Center for Astrophysics; Kitt Peak National Observatory in Tucson,

Arizona; Cerro Tololo Inter-American Observatory in Chile; Cornell University's Arecibo Observatory in Puerto Rico; Los Alamos National Laboratory in New Mexico; and the Canada-France Hawaii Telescope and the Keck Observatory on Mauna Kea, Hawaii. She focused at various times on gamma-ray bursts and binary stars in globular clusters and searched for evidence of water in certain asteroids. Her studies at Harvard were funded by two consecutive grants from the National Science Foundation, in a program called Research Experience for Undergraduates; at Los Alamos a United States Department of Energy grant assisted her. Soderberg identified nine supernovae previously undetected by astronomers; as a college senior she became one of the first human beings to see the most distant supernova recorded up to that point. Her determination and perspicacity impressed the leader of the supernova research group, Brian Schmidt, who told Meredith Gold for the *Portland (Maine) Press Herald* (December 26, 1999), "Discovering supernovae requires one to remain organized and focused for several twenty-hour work days in a row, something Alicia can do as well as any of the team's scientists. Very few students are given opportunities like this. Alicia has made the most of them by proving herself a hard worker, finding her own funding, and asking the right questions of the right people." Guided by Eric R. Wollman, a Bates professor of physics and astronomy, Soderberg wrote a senior thesis entitled "The Efficiency Rate of Type Ia Supernova Searches." She earned a BS degree, magna cum laude, from Bates in 2000.

The following year Soderberg completed an MS degree in applied mathematics at Churchill College, the University of Cambridge in England. She then entered a PhD program in astrophysics at the California Institute of Technology (Caltech) in Pasadena. While there she helped to debunk the commonly held belief in the scientific community that X-ray flashes are ancient gamma-ray bursts or stars exploding in the early universe, detectable only in recent years because of their distance from Earth. She and her thesis adviser, Shrinivas R. Kulkarni, showed that the X-ray flashes originated in modern times in various galaxies. On December 3, 2003, Soderberg observed a cosmic explosion of a kind never before seen, one more powerful than a supernova but fainter than a gamma-ray burst. Until then supernovae and gamma-ray bursts were the only types of cosmic explosions known to accompany the deaths of stars. The burst observed by Soderberg, which became known as GRB031203, appeared in the constellation Puppis, approximately 1.6 billion light-years from Earth. Soderberg earned a PhD degree in 2007. Her doctoral thesis, titled "The Many Facets of Cosmic Explosions," is available as a paperback from Universal Publishers and online booksellers.

## Life's Work

Upon completion of her graduate studies, Soderberg won a NASA Hubble fellowship and the Carnegie-Princeton postdoctoral fellowship. (She won but declined to accept four others, as listed on her curriculum vitae.) She conducted research in association with the Princeton University Department of Astrophysics and the Carnegie Observatories of the Carnegie Institution for Science; the latter include, in California, the Giant Magellan telescope and two Magellan telescopes and, in Chile, the duPont and Swope telescopes. In addition to ground-based telescopes in several continents, she made use of NASA's *Swift* satellite: in January 2008 Neils Gehrels of NASA, who headed the team that managed *Swift*, agreed to Soderberg's request to focus the satellite's instruments on the month-old supernova SN 2007uy in the spiral galaxy NGC 2770. Edo Berger, whom Soderberg was supervising at Princeton, was monitoring the data from *Swift* on January 9, 2008, while Soderberg was giving a lecture when he suddenly noticed an X-ray burst in another part of the galaxy. Berger immediately notified Soderberg, who quickly informed many scientists in the United States and overseas. They in turn commandeered ground-based telescopes and the Hubble space telescope to focus on the new supernova, thus making possible recordings of the entire death of the star. "For years we have dreamed of seeing a star just as it was exploding," Soderberg said during a teleconference with media representatives, as widely quoted on the Internet, "but actually finding one is a once in a lifetime event. This newly born supernova is going to be the Rosetta stone of supernova studies for years to come." (The Rosetta stone made possible the interpretation of Egyptian hieroglyphics in the early nineteenth century.) "We were in the right place, at the right time, with the right telescope on January 9 and witnessed history," Soderberg said, as quoted in *Space Daily* (May 26, 2008). "Thanks to the unique capabilities of the *Swift* satellite and the rapid response of the Gemini telescope we were able to observe a star in the act of dying." Neil Gehrels told Lewis Smith for the *London Times* (May 22, 2008), "This first instance of catching the X-ray signature of stellar death is going to help us fill in a lot of gaps about the properties of massive stars, the birth of neutron stars and black holes, and the impact of supernovae on their environments. We also now know what X-ray pattern to look for. Hopefully we will be able to find many more supernovae at this critical moment" in their evolution.

From 2007 to 2010, Soderberg was a NASA Hubble postdoctoral fellow at the Institute of Theory and Computation at the Harvard-Smithsonian Center for Astrophysics, mapping the diversity of cataclysmic explosions in stars and discovering as rapidly as possible transient bursts of X-rays, radio waves, and light waves and studying their environments, and automating small telescopes. In 2009 she earned the Annie Jump Cannon Award from the American Astronomical Society. Recipients of that honor, which is named for a nineteenth-century American astronomer, are North American female astronomers who earned their doctorates within the last five years. In 2010 she joined the faculty of Harvard University as an assistant professor. Soderberg was awarded a Packard Fellowship in Science and Engineering in 2011 and the 2010 ADVANCE Award from the National Science Foundation and the University of Arizona.

Soderberg told *Current Biography* that on a typical workday, she analyzes data, writes proposals regarding the allocation of time for telescope use, and reads papers in professional journals. She travels often to use large telescopes, give lectures and seminars, and attend conferences. She lives in Cambridge. "I hope to emphasize that a bit of energy and ambition are the key to success in this field," she told *Current Biography*. "You have to be prepared twenty-four hours a day, seven days a week for the unexpected discovery of your career. My ultimate accomplishment would be to disprove the currently accepted notion of 'recipe for success.' A bit of hard work and quick thinking can actually be your winning ticket."

## Further Reading

"Catching the Light of a Baby Supernova." *Space Daily*. SpaceDaily.com, 26 May 2008. Web. 9 Aug. 2012.

Minkel, JR. "Astronomers Witness Supernova's First Moments." *Scientific American*. Nature America, 21 May 2008. Web. 9 Aug. 2012.

Than, Ker. "Supernova Caught Starting to Explode for First Time." *National Geographic News*. National Geographic Society, 21 May 2008. Web. 8 Aug. 2012.

# Spergel, David

## American astrophysicist

**Born**: March 25, 1961; Rochester, New York

"I love exploring the frontiers of science . . . ," the astrophysicist David Spergel told Michael D. Lemonick for *Time Magazine* (2001). "I try to choose projects where the answers will be exciting not only for my colleagues but also for everybody else." Spergel made headlines in 1990 with his theory, which has since been generally accepted, about the shape of the universe, and in 2001 *Time* named him one of the top scientists in the United States. In the same year he won a MacArthur Fellowship—often referred to as the "genius" grant—and watched as his creation for the National Aeronautics and Space Administration (NASA), the Wilkinson Microwave Anisotropy Probe (WMAP), was launched into space. Two years later, after analyzing data from the satellite, Spergel was able to announce to the world his confirmation of the validity of the big bang theory (which holds that the universe was created by a cosmic explosion), as well as his conclusive finding that the universe is 13.7 billion years old. As Michael D. Lemonick wrote, "Even in a field in which the most brilliant minds are inevitably compared with Albert Einstein, Spergel stands out." In addition to his accomplishments as a researcher, Spergel has won praise as a professor at Princeton University. As Scott Tremaine, chairman of the school's Department of Astrophysical Sciences, said for the Princeton University website, "The breadth of [Spergel's] research accomplishments is matched by his enthusiasm for teaching. He has taught—and taught well—almost every undergraduate and graduate course in the department and is one of our most successful and sought-after research supervisors." For the same website, Spergel himself explained: "I think that there is a public view that scientists—particularly those who win awards like the MacArthur—go off and work alone in hidden labs. . . . This is not so. I enjoy working together with students and colleagues. For me, the fun of science is working together to solve interesting questions." Complementing his theoretical work in the field of cosmology, Spergel has gained hands-on experience as a designer of instruments for locating previously undiscovered planets.

## Early Life and Education

David Nathaniel Spergel was born on March 25, 1961, in Rochester, New York. His father, Martin Spergel, is a physics professor at York College of the City University of New York, and his mother, Rochelle Spergel, was a lawyer, now retired. David Spergel's brother, John, is a pediatric immunologist at the University of Pennsylvania, and his sister, Lauren, works in human resources in Lawrenceville, New Jersey. Interested in math, physics, and particularly astronomy even as a child, Spergel joked in an interview with *Current Biography* that he "went into the family business." He attended Princeton University in Princeton, New Jersey, where he received his bachelor's degree in astronomy in 1982. He then enrolled at Harvard University in Cambridge, Massachusetts, where he received his master's degree (1984) and his PhD (1985), both in astronomy. While at Harvard, Spergel spent a year at Oxford University in England as a traveling scholar. After receiving his PhD, Spergel spent an additional year at Harvard as a postdoctoral fellow, serving during that time as a member of the university's Institute for Advanced Study. Meanwhile, in 1987, he returned to Princeton as a professor in the Department of Astrophysical Sciences. He is currently associate chair and director of graduate studies in that department; he is also an associate faculty member of the Department of Mechanical and Aerospace Engineering. Both his research and teaching skills have been lauded by the Princeton administration; the university's president, Shirley M. Tilghman, commented for the Princeton website (October 24, 2001), "From the time of his undergraduate years at Princeton, David Spergel has been an astonishingly bold and creative scholar. . . . He has tackled some of the most difficult and crucial problems in astrophysics and has achieved insights that continue to shape the research agenda in the field. He also has applied his tremendous energy and personal warmth to his teaching, which is greatly appreciated among undergraduates today."

While composing his PhD thesis, "Astrophysical Implications of Weakly-Interacting Massive Particles," Spergel focused his attention on the subject of dark matter, which, according to a widely accepted theory, composes most of the mass in the universe. (Dark matter is matter that cannot be observed by emitted radiation; its gravitational effects on visible objects, such as stars and galaxies, indicate its existence.) This study resulted in his theory that dark matter produces a "wind" of particles that flows against Earth, thus allowing for predictions as to the levels of dark matter that may be detected at different times. (There should be a greater presence in June than in January, for instance.) This proposal is still being investigated by other scientists who specialize in dark matter.

## Life's Work

Spergel told Michael Lemonick that he admires "people who tackle new problems" and who "don't just repeat their PhD research forever." Spergel himself adheres to those standards, as demonstrated when, after his study of dark matter, he attacked a new topic—the shape of the Milky Way. At the 1990 annual meeting of the American Astronomical Society in Crystal City, Virginia, Spergel and his collaborator, Leo Blitz of the University of Maryland at College Park, presented their conclusions on the subject. Their finding was that the Milky Way, the galaxy that contains our solar system, is different from what had been previously believed. Contrary to the widely held belief that the galaxy was spiral-shaped, Blitz and Spergel contended that it is actually shaped like a bar with a spiral at either end. The galaxy contains millions of stars at the center, which are shaped, collectively, like a football, not a globe, as was previously thought. In presenting their findings, as Blitz told Philip J. Hilts for the *New York Times* (January 12, 1990), he and Spergel had found "something of a holy grail for astronomers who study the structure of the Milky Way." Spergel and Blitz's depiction of the galaxy is now the accepted model.

Spergel next focused on the question of cosmic structure—why galaxies form in clusters rather than spreading uniformly through space—a phenomenon that he and a Cambridge University professor, Neil Turok, had researched throughout the late 1980s and early 1990s. At that time they proposed that the clustering occurred because of knots of warped space-time, dubbed "topological defects," which, when questioned by *Current Biography*, Spergel compared to "bubbles in an ice cube. The energy associated with these 'crystal defects' curve and warp

---

**"So as often happens in science, we answer some questions, and we raise and deepen others as our data improves."**

---

space-time." This inventive theory proved to be incorrect: observations from the ground and from satellites did not match the pattern of fluctuations Spergel and Turok had predicted. (Spergel has explained that while there were no errors in his calculations, the calculations were in the service of an erroneous speculation.) In what Michael Lemonick characterized as a "display of intellectual honesty," Spergel admitted that this observational data did not support his model.

At the 1997 American Astronomical Society conference held in Toronto, Canada, Spergel, Blitz (who had relocated to the University of California, Berkeley), Peter Teuben of the University of Maryland, Dap Hartmann of the Harvard-Smithsonian Center for Astrophysics in Massachusetts, and W. Butler Burton of the University of Leiden in the Netherlands presented their findings with regard to another astronomical phenomenon. The topic was the set of mysterious clouds traveling at a high speed across the Milky Way. Observed as early as the 1960s, the clouds have been an enigma for scientists ever since. The major obstacle to unraveling the mystery is that astronomers have been unable to determine the clouds' distance from Earth and, hence, their size. Spergel and his colleagues suggested at the convention that the clouds are more than a million light-years away from our galaxy and that their movement can be attributed to the gravity of the Andromeda galaxy and the Milky Way. They also proposed that the gas that composes the clouds is a remnant of the primordial gas that coalesced to form the entire Local Group of galaxies, the group that contains the Milky Way and the Andromeda galaxy, among other objects. This conclusion was supported by a computer simulation of the formation of the Local Group but was not accepted as a definitive resolution of the mystery surrounding the clouds. Spergel's theory is considered viable, but scientists are still pursuing other possible explanations as well. If the theory is correct, though, the presence of relic gases would sustain the present rate of star formation for billions of years to come.

### The Wilkinson Microwave Anisotropy Probe

His earlier display of ethics, Spergel has said, earned him an invitation in 1994 to work with scientists at NASA as the principal theorist on the design team for the Wilkinson Microwave Anisotropy Probe (WMAP). WMAP was designed to map the oldest radiation (or energy radiated in wave or particle form) in the universe by recording cosmic radio waves. Although, as a theoretical astrophysicist, Spergel had no previous practical experience in satellite creation, he threw himself into the work, often rising at seven in the morning and working until two in the morning. After eight years of labor, the 1,800-pound WMAP satellite was launched from Cape Canaveral, Florida, in June 2001. Following the transmission of the WMAP data, Spergel and his team revealed their findings in early 2003. The robotic probe had created the clearest pictures of radiation ever taken, which, when examined, allowed Spergel and his colleagues to fix the exact age of the universe at 13.7 billion years—a discovery that marked the first time ever that the age of the universe had been determined conclusively. These findings also provided clear evidence of the big bang, only a theory until confirmed by WMAP data, and determined that the big bang occurred 14 billion years ago. In addition, the data revealed that only four percent of the universe is made of atoms, whereas the other 96 percent is composed of invisible dark matter. Spergel commented to Ira Flatow on the National Public Radio program *Talk of the Nation* (February 21, 2003), "So as often happens in science, we answer some questions, and we raise and deepen others as our data improves." Spergel called his work on the WMAP one of the most satisfying experiences of his life.

Due largely to his contribution to WMAP, Spergel was awarded a 2001 MacArthur Fellowship by the John D. and Catherine T. MacArthur Foundation. The fellowships, which consist of no-strings-attached monetary awards distributed over a five-year period (Spergel received $500,000), are given to those individuals who, according to the MacArthur Foundation website, demonstrate "extraordinary originality and dedication in their creative pursuits and a marked capacity for self-direction" as well as "exceptional creativity, promise for important future advances based on a track record of significant accomplishment, and potential for the fellowship to facilitate subsequent creative work." "The MacArthur Fellowship is both a wonderful opportunity and honor," Spergel said for the Princeton University website (October 24, 2001). "The fellowship will help me juggle the challenges of research, teaching and three young children."

After the success of the WMAP satellite, Spergel again went to work for NASA on designing a second spacecraft intended to discover Earthlike planets in other solar systems. Spergel knew little about optics or telescope design when he began the project but, as he told Michael Lemonick, "I got a book and taught myself optics." He then proposed the creation of a telescope with an innovative lens design that could detect a dim planet by blocking the light of brighter stars. NASA hopes to launch the telescope, called the Terrestrial Planet Finder, in approximately ten years.

Spergel has received a variety of awards during the course of his career and has published numerous articles in such publications as the *Astrophysical Journal*, *Nature*, and *Physics Letters*. He is currently chair of NASA's Origins Subcommittee, which oversees the Hubble Space Telescope, the James Webb Space Telescope, and NASA's program of searching for undiscovered planets. The committee advises NASA on long-term goals rather than technical aspects of space exploration. He is also involved in a project at the Atacama Cosmology Telescope in Chile, studying cosmic microwaves. In 2010, Spergel shared the Shaw Prize in Astronomy with Charles Bennett and Lyman Page Jr., for "their leadership of the WMAP experiment, which has enabled precise determinations of the fundamental parameters . . . of the universe," according the prize's webpage.

Since August 26, 1990, Spergel has been married to Laura H. Kahn, a physician, with whom he has three children: Julian, Sarah, and Joshua. In his free time, Spergel is an avid skier and bicyclist.

## Further Reading

Flatow, Ira. "WMAP Project and How It's Answering Questions About the Universe." *Talk of the Nation/Science Friday.* National Public Radio, 21 Feb.2003. Web. 9 Aug. 2012.

Folger, Tim. "Cosmologist David Spergel-Decoder of the Cosmos." *Discover Magazine* May 2003. Print.

Lemonick, Michael D. "Astrophysics: Mr. Universe." *TIME Magazine.* Time, 20 Aug. 2001. Web. 9 Aug. 2012.

# Spiropulu, Maria

## Greek physicist

**Born**: 1969; Greece

"We are very close to a new reality," the physicist Maria Spiropulu told Peter Gorner for the *Chicago Tribune* (February 18, 2002). "We imagine space and time as static entities and we solve equations as a function of space and time. But we're learning that at the very large scale—the cosmos—and at the very teeny scale of particles, space and time are dynamic, constantly changing. What is happening at those scales we can't explain. So we have to wonder: Do these scales hold some extra dimensions?" Spiropulu's research focuses on that question; by making particles collide and then observing the results, she hopes to discover the existence of dimensions outside both three-dimensional space and time. A former fellow at the Enrico Fermi Institute at the University of Chicago, Spiropulu currently works at CERN, the world's largest particle-physics laboratory, in Geneva, Switzerland. She is a member of a new generation of physicists who are challenging the traditional ways of understanding the material world. "Physics is not my job; it is my life," she told Dennis Overbye for the *New York Times* (September 30, 2003). "The world is what you measure."

## Early Life and Education

Maria Spiropulu was born in the fall of 1969 in Greece and raised in Kleisoura, a village of roughly one thousand residents located in the northern mountains of Greece. Her younger brother, Markos, a world rowing champion at the age of fourteen, is currently a yachting instructor and professional skipper in Athens. Spiropulu's father was a businessman; her mother was a teacher of fashion design. As a child, Spiropulu had a passion for reading and a taste for music and theater. She was also fascinated by the way things work and loved to take objects apart and put them back together. Early on she began conducting scientific experiments; in one, carried out when she was seven, she made a pair of Styrofoam wings and fastened them to the arms of one cousin after another, each of whom jumped from the top of a garden wall at her command. When the would-be aerialists landed on their faces, she told them to flap harder the next time they jumped.

Spiropulu excelled not only in the sciences and mathematics but also in philosophy, history, and literature. Her teachers encouraged her to consider studying philology and law. By the time she was fourteen, though, she knew she wanted to specialize in physics. In high school she developed an interest in particle physics. As an undergraduate at Aristotle University of Thessaloniki in Greece, she found herself, in her own words, in a state of "creative confusion." Among other subjects, she studied material science, surface physics, and dosimetry, defined as "a system of therapeutics which uses few remedies, mostly alkaloids, and gives them in doses fixed by certain rules." After she graduated from the university with a bachelor's degree in physics in 1993, she borrowed $500 from her father and moved to Geneva, Switzerland. There, she got a summer job as a technical assistant at CERN (the acronym for Conseil Européen pour la Recherche Nucléaire, or European Laboratory for Particle Physics), the world's leading institution devoted to the study of fundamental particles.

During that summer she resolved to build a career in particle physics, and she set her sights on investigating proton-antiproton collisions at the Fermi National Accelerator Laboratory (Fermilab) in Batvia, Illinois. Fermilab is the home of the Tevatron, a circular structure in which particles are accelerated to an enormously high speed: only a tiny fraction of a second slower than the speed of light, which travels at 186,000 miles per second, or 669.6 billion miles per hour. With a circumference of 6.5 kilometers, or slightly over four miles, the Tevatron was the world's largest particle accelerator at the time (the Large Hadron Collider, completed in 2008, has since surpassed it). Connected with the Tevatron are two so-called detector complexes—the Collider Detector Facility and the D0 Detector Facility, each of which is three stories tall; the detectors determine the presence of all the elementary particles and fragments of particles produced in particle collisions.

At the end of that summer, Spiropulu enrolled at Harvard University in Cambridge, Massachusetts, as a graduate student in particle physics. During much of the next half-dozen years, she served an apprenticeship at Fermilab investigating particle collisions. To facilitate her project, she developed a mathematical-analysis tool that searched the accelerator's data for evidence of supersymmetry.

According to the theory of supersymmetry, all known particles have associated "shadow partners." The nature of the shadow partner depends upon the type of particle with which it is associated: if

---

**"Physics is not my job; it is my life."**

---

the particle is of the type categorized as a boson, which transmits a force, its shadow partner (also called a partner particle or superpartner) will be of the type categorized as a fermion, which makes up matter; conversely, the counterpart of a particle of the fermion type will be a particle of the boson type. Among fermions, the superpartner of an electron is called a selectron; the superpartner of a muon is a smuon; that of a quark, a squark, and so on. (The name of a fermion's superpartner is generated by adding "s" to the name of the particle.) Among bosons, the superpartner of a photon is a photino; that of a gluon, a gluino; that of a Higgs, a Higgsino, and so on. (The name of a boson's superpartner is formed by substituting "ino" for "on" or simply adding "ino.") In addition, the particle/superparticle connection involves the highly theoretical concept of spin, which in particle physics is defined as the intrinsic angular momentum of a particle (and has nothing to do with anything actually spinning). According to the theory of supersymmetry, the spins of a particle and its associated superparticle differ by half a unit. Thus, the spin of a Higgs is zero, and the spin of a Higgsino is one-half; the spin of an electron is one-half, and the spin of a selectron, zero. Spiropulu searched for evidence of gluinos and squarks, the superpartners of gluons and quarks, respectively. The fact that she did not find them did not detract from her research, because, as the doctoral judges determined, her methods and calculations were correct. "Everybody is entitled to their own opinion, but they're not entitled to their own facts," she remarked to Dennis Overbye. "The data is the data." Spiropulu earned her PhD in 2000; her dissertation bears the title "A Blind Search for Supersymmetry in p pbar Collisions at 1.8 TeV Using the Missing Energy plus Multijet Channel." In the same year she became an Enrico Fermi Fellow at the Enrico Fermi Institute at the University of Chicago. Concurrently, she continued to study particle collisions at Fermilab.

## Life's Work

At the 2002 annual convention of the American Association for the Advancement of Science, held in Boston, Massachusetts, Spiropulu presented a paper on experiments that had led her to believe that a new understanding of reality would emerge by 2005. "Reality, to us, is a world where events happen over time within a three-dimensional space, but the way we think about things is about to completely change," she predicted at the convention, as quoted by Peter Gorner in the *Chicago Tribune* (February 18, 2002). According to traditional physics, humans exist in four dimensions: one time direction and three spacial dimensions (left-right, forward-backward, and up-down; a line is one-dimensional, a square is two-dimensional, and a cube is three-dimensional). In recent years, physicists have become increasingly interested in superstring theory, which aims to shed light on such phenomena as gravity and the quantum nature of light, which traditional physics cannot explain. String theory attempts to unravel the mysteries of space, time, matter, and energy by positing the existence of tightly bundled, vibrating "strings" of incredible tininess (about a millionth of a billionth of a billionth of a billionth of a centimeter). The theory makes mathematical sense only if there are ten dimensions rather than four. In order to explain the fact that humans experience only four dimensions, string theorists suggest that the extra dimensions are rolled up into tiny balls less than a trillionth the size of elementary particles. The existence of extra dimensions might also help explain why gravity is so weak compared with the other three forces of nature: electromagnetism, the strong nuclear force, and the weak nuclear force (the latter two of which act only within the nucleus of the atom); if other dimensions exist, then gravity appears weak because it must be transmitted through those dimensions.

In order to discover whether extra dimensions really exist, Spiropulu and her colleagues at Fermilab conduct experiments in high-energy particle collision. Using the Tevatron, they force protons and their antimatter mirror particles, antiprotons, to collide with one another. (Antimatter mirror particles are different from shadow partners.) Theoretically, these collisions should produce gravitons, a hypothetical particle of force that carries gravity. (Physicists have not yet detected gravitons; they believe they exist because particles associated with the other three forces have been detected.) According to traditional physics, gravitons are massless and undetectable. The forced particle collisions in Spiropulu's experiments produce less energy than mathematically predicted, thus suggesting that some of the particles might have been absorbed into extra dimensions. "We do have things that give us the right to think we are seeing a little hint of new physics," she told Dennis Overbye. "It's an educated hope. It's not wishful thinking."

From 2004 until 2011, Spiropulu conducted her research at CERN. She became a fellow of the American Association for the Advancement of Science in 2008. In 2009, she accepted a position as associate professor of physics at the California Institute of Technology (Caltech); in 2012, she became a professor. In her spare time, Spiropulu enjoys dancing the tango, kickboxing, and playing the drums. While working toward her PhD, she served as a drummer for Fermilab's resident band, Drug Sniffing Dogs.

## Further Reading

Butler, Sharon. "University of Chicago Particle Physicist Maria Spiropulu is…" *Chicago Tribune* 11 July 2002: T1. Print.

Johnson, George. "New Generation of Physicists Sustains a Permanent Revolution." *New York Times* 20 June 2000: F1. Print.

Kotulak, Ronald. "Seriously Weird Science." *Chicago Tribune* 11 January 2004. Print.

Overbye, Dennis. "Maria Spiropulu; Other Dimensions? She's in Pursuit." *New York Times* 30 Sept. 2003: F1. Print.

# Turin, Luca

## British biophysicist

**Born**: November 20, 1953; Beirut, Lebanon

Luca Turin "is not just a fragrance chemist," as Vicki Allan wrote for the Sydney *Sunday Herald* (May 21, 2006). "He is one of those unorthodox scientists—originally a biophysicist—who has transcended disciplines and dared to come up with a big theory." That "big theory" proposes that smell is dependent not on the shape of the molecules of the substance emitting an odor, as most scientists believe, but on the speed of vibration of the bonds of that substance's molecules. When Turin introduced his theory in 1995, specialists in his field immediately labeled him a renegade of olfactory science, and that label has remained: to date, his theory has not been corroborated by any other scientist. At the same time, it has never been disproved, and some of Turin's findings cannot be explained by other theories. Turin is the subject of the biography *The Emperor of Scent: A Story of Perfume, Obsession, and the Last Mystery of the Senses* (2002), by Chandler Burr (who currently critiques fragrance for the *New York Times*). Interested since early childhood in the nature of odors, Turin has taught at University College London, a division of the University of London in England; acted as a consultant for major perfume producers; and published three books about perfume: *Parfums: Le Guide*, written in French; *The Secret of Scent: Adventures in Perfume and the Science of Smell*; and *Perfumes: The Guide*, which is a revised, English version of his first book and was written with his wife, Tania Sanchez. In a review of *The Secret of Scent* for the *New York Times* (December 3, 2006), John Lanchester described Turin as "a brilliantly readable perfume critic" and wrote that he "has an extraordinary gift for writing about smell." Turin is the chief technological officer for Flexitral, a Chantilly, Virginia-based company that uses Turin's odor-prediction technology to develop fragrances far more quickly and less expensively than is possible with traditional methods.

The scent industry accounts for about $14 billion in sales annually. In addition to perfumes (currently, 600 to 1,000 new scents are introduced every year), scents are added to a large array of products, among them soaps and detergents, deodorants and other toiletries, room fresheners, candies, and even new cars. With the costs of many natural scents, such as sandalwood, increasing exponentially in recent years, the quest for artificially created chemical scents has intensified. Thus, potentially, Turin's work could revolutionize the industry.

## Early Life and Education

Luca Turin was born on November 20, 1953, in Beirut, Lebanon, to Adela Mandelli and Duccio Turin, a leading expert in the field known as building or construction economics. Both of his parents, who met in Italy, were Argentineans of Italian descent. Mandelli was a designer for the couturier Lanvin Castillo during the 1950s and 1960s, while Duccio Turin worked as an architect and town planner with the United Nations Refugee Welfare Association; his assignments included designing refugee camps for Palestinians. Soon after Luca Turin's birth, the family settled in Paris, France, where Duccio Turin joined the staff of the Centre Scientifique et Technique du Bâtiment, which is concerned with the construction of human habitation. In 1957 he resumed working for the United Nations and moved with his family to Geneva, Switzerland, where he and his wife soon divorced.

Luca was extremely unhappy in Geneva, so his mother returned to Paris, where she became the art director for an advertising agency. Meanwhile, at age four, Luca had learned to read, absorbing the reading lessons that his mother was giving to his Spanish nanny. When he was seven his mother rented a beach house on the Cote d'Azur (a famous resort area known in English as the French Riviera). "The moment we arrived in this strange place he set about systematically analyzing the smell of the thyme that grew wild everywhere," his mother recalled to Chandler Burr. Back in Paris, Turin became enamored with science and enjoyed visiting the Palais de la Découverte, a science museum. According to Burr, as a youth "he was famous for boring everyone to death with useless, disconnected facts, like the distance between the earth and the moon in Egyptian cubits."

Turin and his mother later moved to Milan, Italy, where Mandelli had taken a job as the manager of product and image design with Upim, a department store. Later, in 1974, she founded a feminist publishing house for children's books, called Dalla Parte delle Bambine. (It currently operates under the name Du Côté des Filles.)

Earlier, in 1966, Duccio Turin had joined the faculty of the Bartlett School of Architecture at University College London. In 1976 he served as a deputy secretary-general for the United Nations Habitat Conference in Vancouver, Canada. Within days of its conclusion, he was killed in a car crash. The book *Construction and Economic Development: Planning of Human Settlements: Essays in Memory of Duccio Turin*, edited by Otto H. Koenigsberger, was published in 1978.

## Life's Work

Luca Turin earned a PhD in physiology and biophysics at University College London in 1982. After his graduation he moved to the south of France, near Nice, to conduct research in marine biology at the Villefranche Marine Station as an employee of the Centre National de la Recherche Scientifique. During his free time he often hunted for perfumes in retail shops. At the suggestion of someone he met on one such excursion, he visited Claudine Pillaud, a purveyor of rare perfumes in Menton. Pillaud sold perfumes only to customers whom she liked. At their first meeting Turin impressed her when he correctly identified the principal note in the perfume Bellodgia as carnation. (The ingredients of any perfume are divided into three categories called notes, each of which evaporates at a different rate.) Pillaud sold him a one-ounce tester bottle of Diorama, the first fragrance Christian Dior marketed after World War II, which contained perfume made according to its original formula. Like many other perfumes, in the early 1980s Diorama was being manufactured with cheaper materials; the materials for the original formulations were not only rare but very expensive. "The financial temptation to dilute perfumes is almost irresistible," Turin explained to Burr. Referring to the oil of the agarwood tree, which is used as a base for perfumes, he continued, "I mean, if you can make your oudh go twice as far. . . . Instead of using ten kilos you use five, and given how much this stuff costs per gram, the temptation to futz is just irresistible. It's why Guerlain and Chanel and Jean Patou . . . are so great. Not necessarily expensive perfumes. Just great ones that are never, ever diluted. No tricks, no cheating, no cutting corners. But today all bets are off. When the big fragrance firms take L'Air du Temps and wreck it by having an accountant redraw the formula to take out the expensive ingredients and substitute cheap ones, what they are doing, among other things, is depriving thousands

> **"Smell must be written in molecules, but no one has been able to read that language. It's a bit like an archaeologist trying to decipher hieroglyphics."**

of people throughout the world of the thrill of the memories that are infused with L'Air du Temps, because unless it is the same smell, it won't trigger [a memory]. The pale new reflection may be intellectually, objectively, a reflection . . . but your brain stem isn't electrified. Memory isn't triggered." Thereafter, during that period, Turin spent a great deal of his money on collector perfumes.

In 1990 an American friend of Turin's told him that the United States Navy was interested in researching and developing sensors to detect odors from enemy submarines in order to trail them, and that the Department of Defense was going to award grants for such studies. Turin applied for a grant, but before he had heard back from the Department of Defense, he was asked to leave the Centre National de Recherche Scientifique, after inadvertently implicating a famed French scientist in a case of fraud. Turin then moved to North Carolina, where he had gotten a position at a branch of the National Institutes of Health (NIH). Soon afterward the funding for the NIH project on which he was working ran out, and he returned to Paris. Unemployed, he began to write perfume reviews, both for his own amusement and as a source of income. The first three fragrances he critiqued were Grace a Rive Gauche, Apres l'Ondee, and Vetiver. Of the first he wrote in *Parfums: Le Guide* (as translated into English), "Thanks to Rive Gauche, mortals can at last know the scent of the goddess Diana's bath soap. A true emblem of the 70s, this sumptuous reinterpretation of the innovative metallic note found in the less fortunate Calandre (Paco Rabanne) belongs to the uncrowded category of sculpture-perfumes. Its seamless silvery form, initially hidden by white, powdery notes, soon pierces the clouds and gains height by the hour."

Hermes agreed to publish Turin's collection of reviews, with the title *Parfums: Le Guide*, in 1992. Now out of print (in the French version), it became the bestselling perfume guide in France and caught the attention of the world's seven major perfume producers: International Flavors and Fragrances, Givaudan Roure, Quest International, Firmenich, Haarmann & Reimer, Dragoco, and Takasago. At the invitation of Quest International, Turin visited the company's labs to see how their perfumes were made and voice his opinions on various fragrances.

Still unemployed, Turin moved to Moscow, Russia, after accepting an offer of a visiting professorship from the Moscow State University Institute of Molecular Biology. Soon tiring of that job, he contacted a former professor of his at University College London and secured a teaching job there in 1993. At the same time the US Navy notified Turin that he had won the $60,000 in research funds that he had requested. One day, while perusing a copy of the *Review of Scientific Instruments*, he read about an electron-tunneling spectroscope, a device that measures the vibrations of the molecular bonds of particular substances and thereby indicates the identity of the substances. The mode of operation of the spectroscope led Turin to wonder if, with regard to detecting scents, the nose might work in a similar manner, by somehow distinguishing among molecular vibrations in the act that we call smelling. (An earlier scientist, Malcolm Dyson, suggested in 1938 that the nose operated like a gas chromatograph—an idea that was instantly and universally dismissed.) The accepted theory of the mechanism of the sense of smell centers on the shape of given molecules. Turin questioned the validity of that theory; as he told an interviewer for the *London Daily Mirror* (November 27, 1995), "Smell must be written in molecules, but no one has been able to read that language. It's a bit like an archaeologist trying to decipher hieroglyphics."

Turin acquired the use of a Quest International lab, where he began to test his theory by comparing the scents and vibrations of borane (a compound, not found in nature, composed of hydrogen and boron) and the element sulfur. Turin showed that molecules of borane and sulfur differ in shape but have similar odors and similar vibration speeds; thus, he concluded, vibration must determine the smell. He also found that acetophenone (a chemical, found in several fruits and elsewhere, that is used in the manufacture of fragrances) and deuterated acetophenone (acetophenone to which deuterium, an isotope of hydrogen, has been added) have dissimilar odors and different molecular vibrations yet are the same shape and size. When he submitted a paper describing his hypothesis and experimental results to the prestigious scientific journal *Nature*, the editor rejected it, declaring in a letter, "The major body of the paper is an extremely rambling case-by-case discussion of specific molecules and their shapes. I find this quite unconvincing and virtually irrelevant." In 1996 a far less influential journal, *Chemical Senses*, published Turin's paper.

That year Turin became an outside evaluator for Quest International's fragrances. The company, which was in the process of negotiating a contract to produce Dior fragrances, fired him after he told the French beauty magazine *Votre Beaute* that Dior had unfavorably altered its perfumes by using cheaper ingredients. Quest kept him on briefly, however, to experiment with scent prediction—that is, predicting how a newly created perfume ingredient will smell—and he developed a computer algorithm to analyze the potential scent of a molecule based on its vibrations. Traditionally, in creating new scents, other producers relied on a system of trial and error: technicians had to build molecules, which experts then sniffed, and often, as many as 2,000 molecules were constructed before one or two were deemed satisfactory. That labor-intensive process was extremely costly.

After Turin's association with Quest ended, representatives of another perfume manufacturer, Takasago, met with him once, to talk about developing a machine to create musk-scented molecules, then opted not to deal with him. Burr claimed that the major firms' rejection of Turin stemmed from reasons unrelated to the merits of his approach.

"We want to believe that science is dispassionate, objective, and (for those who don't have use for a theological god), omniscient," Burr wrote. "We want to believe that every idea that merits attention is given it. That the good ideas are kept, the bad ones discarded, the industrious rise, the lazy sink, and the hard work and honest data are rewarded. This isn't real. Perhaps unfortunately, perhaps not. Scientists are human. Vested interests beat out new ideas. Egos smother creativity. Personalities clash. Corruption is as common as the survival instinct."

In 2001 a newly launched firm, Flexitral, hired Turin as its chief technological officer. According to its website, "The strategy followed by Flexitral since its creation late in 2001 is to identify the fragrance and flavor industry's most pressing needs and answer them by rational odorant design. Flexitral has developed methods to predict the smell of a molecule before it is synthesized. . . . Products were reached after synthetic programs of less than ten molecules." Within the first year, with only $1 million in investments, the company produced two new scents: Acitral and Lioral, which are redolent of lemon and lily-of-the-valley, respectively. Flexitral's products currently include, in addition to Acitral B and Lioral, the jasmine-tinged floral Jasphene, the violet-like Neoctin, the carnation-fragranced Nugenol, the black-currant-like Ribescone, the coumarin-like Coumane, and two rosy-citrus scents, called Tricitrol and Trinalool.

Turin has acknowledged that his theory is more a rebuttal of current theories than an empirically sound explanation of the mechanism of smell. He told Mick O'Hare for *New Scientist* (November 18, 2006), "Every scientist has two jobs—demolition and construction. I've done a good demolition job. . . . Demolition is OK, but what do you construct instead? Empirically the easiest test of vibrational theory would be if different isotopes of the same element smelled different. For example, if you replace hydrogen with heavy hydrogen, the molecular shape is identical but the vibration is different. So do isotopes smell different or not? There's no consensus."

Turin's second book is *The Secret of Scent: Adventures in Perfume and the Science of Smell* (2006). Although most critics acknowledged Turin's ability to write lyrically about perfumes, they often found that his scientific explanations of vibration theory and other matters were impenetrable. The book "is an admirably sustained feat of evocation," Alex Butterworth wrote for the *London Observer* (May 14, 2006), "and if the cumulative effect is a little heady, there is real poetry here to stimulate our jaded perception." In a review for *Time Out* (May 17, 2006), Leila Dewji wrote, "From the outside, the world of perfume seems glamorous, exotic, and sexy; but once you reduce it to molecular structures, atomic weights, and the Nobel-winning findings of lesser-known Croatian chemists, it all becomes rather laborious and dull. There are pages of incomprehensible diagrams, and it's not always clear which part of the text they're illustrating." The English edition of *Parfums: Le Guide*, which Turin rewrote with his wife, Tania Sanchez, was published in 2008.

Turin has said that fragrance is so important to him that if he had not met a woman who shared his passion for it, he would have stayed unmarried. He and his wife live in London.

## Further Reading

Colquhoun, Kate. "Scents and Acute Sensibilities." *Daily Telegraph* 17 June 2006: 6. Print.

Hill, Stephen. "Sniff 'n' Shake." *New Scientist* 2115 (3 Jan. 1998): 3434. Print.

Lanchester, John. "Odor Decoder." *New York Times Sunday Book Review* 3 Dec. 2006: 32. Print.

# Appendixes

# Historical Biographies

## Niels Bohr

Danish physicist

*Bohr discovered the fundamental structure and character of the atom, its components, and how they interact. For this discovery, he won the Nobel Prize in Physics in 1922. He also made significant contributions to the understanding of how quantum and classical physics unify as a single philosophy in his principle of complementarity.*

**Areas of Achievement**: Physics, mathematics

**Born**: October 7, 1885; Copenhagen, Denmark

**Died**: November 18, 1962; Copenhagen, Denmark

### Early Life

Niels Bohr (nihls bohr) was born into a family environment that invited genius; his father, Christian Bohr, was a professor of physiology at the University of Copenhagen, and his mother, née Ellen Adler, came from a family of eminent Danish educators. Bohr's younger brother, Harald, would become a professor of mathematics.

Bohr attended Gammelholm Grammar School and entered the University of Copenhagen in 1903. He studied under the tutelage of C. Christiansen, a prominent physicist and an original, creative educator. At the University of Copenhagen, Bohr took his master's degree in physics in 1909 and his Ph.D. in 1911. He published his first scientific work in 1908. The opportunity arose as a result of a prize offered to the individual who solved an investigation of surface tension by means of oscillating fluid jets. He won the gold medal, and his piece appeared in the *Transactions of the Royal Society*.

In the fall of 1911, Bohr studied abroad at the University of Cambridge, pursuing largely theoretical studies under Sir Joseph John Thomson. However, he had not been at Cambridge long before he realized that the true frontier work in theoretical physics was occurring at the laboratories of Nobel laureate (1908) Ernest Rutherford at the nearby University of Manchester. Bohr also was drawn by Rutherford's dynamic personality. Before Bohr arrived at Manchester in the spring of 1912, Rutherford had already deduced the structure of the atom experimentally, although the concept's theoretical foundation still held significant flaws. Bohr, however, was about to uncover an idea that would forever change the face of physics and the very concept of the physical world itself.

## Life's Work

As Bohr pondered the beauty of the emerging picture of the atomic structure, he, like Rutherford, was perplexed by the evident contradictions in theory. Rutherford's atomic model held that the atom was made of a very dense, positively charged central core surrounded by a cloud of negatively charged particles. Yet, based entirely on Newtonian, classical physics, such a structure could not exist, becoming unstable and falling apart. It was no wonder that Rutherford's peers held that the theory was fatally flawed. Yet Bohr had an almost heroic faith in Rutherford's insight and experimental proficiency, so he stubbornly clung to the idea that the experimental evidence had only to be matched with the appropriate theory.

In 1911, the infant science called quantum mechanics had found few applications. Bohr recognized that by linking the statistical methods of quantum mechanics with an invariant number called Planck's constant, for Max Planck, he could theoretically vindicate Rutherford's experimental evidence. Bohr reasoned that energy from the atom, emitted only in well-defined energy levels, was related to electrons falling or rising into stable orbits around the nucleus, a concept somewhat alien to Isaac Newton's classical notions of cause and effect. Using Planck's constant, he was able to derive the calculations necessary to describe the stability and transitions of the electrons, thus defining precisely the nature of the atom itself. The results of his work were ultimately verified by experimental evidence. He published these results in 1913 and for them would win the Nobel Prize in Physics nine years later.

From this work, Bohr reasoned that the model for the atom described in quantum terms must join smoothly with classical physics when the dimensions become larger than atomic size. This logic vindicated Rutherford's physical, experimental evidence as essentially correct. Yet in a larger sense, the idea enabled a philosophical justification of using the quantum set of scientific rules to describe the atomic world and the classical set to describe the larger universe. He called this fusion of ideas the principle of correspondence.

In 1916, Bohr returned to Copenhagen an acclaimed physicist. By 1920, he was named the director of the Institute of Theoretical Physics at Copenhagen. It would later be renamed the Niels Bohr Institute. By this time, Bohr had continued his investigations and uncovered evidence that the active properties, degree of stability, and character of all matter itself were largely dependent on the arrangement of the electron shells of the elements. Yet all the ongoing descriptions of the atomic character were being almost wholly defined in quantum terms, many being derived from a form of statistical probability. Albert Einstein was dissatisfied with this state of affairs, which violated his sense of universal simplicity, and stated, "God does not play dice with the universe."

Bohr recognized that this debate was threatening the very foundation of theoretical physics. He used the example of two experiments to deliver what he called the principle of complementarity. The experiments he referenced were ones that unambiguously showed the electron to be a wave form and another that showed it in equally definitive terms to be a particle. He said that one observation necessarily excluded viewing results that only the other could obtain and vice versa but that one did not necessarily disprove the other. Together, however, the concepts were complementary proofs of each other. It was all a matter of philosophy, as was the foundation of science itself. The principle of complementarity stood as a brilliant turning point in physics. With it, the dominance of classical physics was ended and the new insights of the surreal quantum world were engendered.

Bohr's reputation attracted physicists the world over to Copenhagen to study and discuss the direction of modern physics. Bohr and Einstein frequently debated the consequences of quantum mechanics and the effects of this science on the perception of causality.

Denmark was taken by the Nazi storm troopers in 1940. Bohr was an outspoken anti-Nazi, and his mother was Jewish. Under the threat of imminent arrest, Bohr and his family escaped Copenhagen in 1943 on a fishing boat to Sweden. They eventually traveled to the United States. Bohr worked with the Allies' most influential physicists under the direction of J. Robert Oppenheimer, and they succeeded in building the nuclear bomb that would end the war. Indeed, it was Bohr who first predicted that the isotope uranium 235 would be the element of choice for a nuclear weapon.

After the war, the golden age of physics was irrevocably ended for Bohr. He set about at once to convince both Franklin D. Roosevelt and Winston Churchill of the immediate need to control nuclear weapons. Failing this, he helped establish the First International Conference on the Peaceful Uses of Atomic Energy, eventually winning the first Atoms for Peace Award in the United States (1957). Bohr died in Copenhagen on November 18, 1962.

## Significance

Using the concept of complementarity, Bohr forged the link between quantum mechanics and classical physics. This linkage enabled science to move deliberately ahead into subatomic research using a philosophy that was both unique to the atom's peculiar interior and relative, in a complementary sense, to a larger world. Bohr was a restless scientist who believed that there was an innate unity to many aspects of the physical world, expressed in the abstractions of complementarity, yet he was also practically oriented.

Bohr's creation and forceful leadership of the Institute of Theoretical Physics in Copenhagen served as a wellspring of emerging knowledge about the atom to scientists worldwide. Yet to Bohr it also represented an undisguised empire, directing the pace and direction of atomic science. From Copenhagen, he would personally coordinate and trace the direction of theoretical physics as his institute became the center of world attention to this strange, new science.

His empire would both collapse and race out of his control with the ascension of Adolf Hitler's war machine and his consequent contributions to the development of the atomic bomb by the United States at Alamogordo, New Mexico. He was one of the first scientists to grasp fully the aggregate implications of the bomb to humanity, ex post facto. When Bohr returned to postwar Copenhagen, the impetus of theoretical physics had shifted to the United States. His own energy would be devoted to attempting somehow to repair the damage or at least to slow the proliferation of the nuclear bomb. Yet his influence on modern physics will profoundly overshadow the misfortunes of war and the misapplication of science. The ability of one person logically to unite the seemingly disparate worlds of quantum and classical physics in a philosophical, mathematical unity called complementarity stands as one of the most important and astonishing intellectual triumphs of science.

**Dennis Chamberland**

# Marie and Pierre Curie

## French chemists

*The Curies made some of the most significant scientific advancements in the modern age with the discovery of radium and other radioactive elements. For this work, the Curies and scientist Henri Becquerel received the 1903 Nobel Prize in Physics. Marie alone would win the 1911 Nobel Prize in Chemistry, one of the few individuals ever to win two such awards.*

---

**Areas of Achievement**: Chemistry, physics

**Born**: November 7, 1867; Warsaw, Poland

**Died**: July 4, 1934; Sancellemoz, near Sallanches, France

---

### Early Lives

Pierre Curie (pyehr kyoo-ree), the son of a doctor, Eugène Curie, and his wife, Sophie-Claire, was born in Paris. He was taught at home by his father. At the age of fourteen, he developed an avid interest in mathematics. He studied at the Sorbonne, from which he received his bachelor of science degree, and later became a laboratory assistant there. In 1882, he was appointed supervisor at the École de Physique et de Chemie Industrielle in Paris, as well as serving there as a professor from 1894 until 1904. Also in 1894, Pierre met Marya Skłodowska.

Skłodowska, later known to the world as Marie Curie, was born in Warsaw, part of the Russian-controlled territories of Poland. Marya, or Manya, as she became known, was the youngest of five children in the Skłodowski family. Her father, a professor at a Russian gymnasium (the equivalent of a high school) in Warsaw, encouraged in his children an appreciation for knowledge and learning that they would carry throughout their lives.

Manya learned to read by the age of four and soon exhibited an exceptional aptitude for science and mathematics. In 1883, when she was sixteen, she won a gold medal for academic achievement on her graduation from her Warsaw gymnasium. After graduation, she went to work as a tutor and governess in Warsaw and other areas in Poland to help support her family and to finance her eldest sister's studies at the Sorbonne. During this period, she also participated as an instructor in an underground university in Warsaw, an unofficial school for Polish students that taught science, mathematics, and Polish history, subjects forbidden by the Russian forces occupying the country.

In November of 1891, Manya moved to France with her eldest sister and enrolled at the Sorbonne as a physics student. Once in Paris, she changed her first name to Marie, the French version of Marya. Living a spartan existence, Marie immersed herself in her studies and, in 1893, earned a license of physical sciences (the equivalent of a modern B.S. in physics), placing first in her class. The following year, she was second in her class and earned a license in mathematical sciences. After graduation, Marie was introduced to Pierre Curie, who by this time had become known for his studies in the properties of different types of crystals. The two soon discovered that they shared a passion for science and advanced research. The following year, on July 25, 1895, the two scientists were married, beginning a partnership that would change the face of modern science in the twentieth century.

## Lives' Work

The Curies' work on the nature and properties of radioactive materials had the same kind of humble beginnings that are characteristic of many scientific milestones in the modern age. They were, in fact, as much a result of an accident someone else's as of the hard work and imagination of the young couple themselves. The French scientist Antoine-Henri Becquerel, studying the possibility that a variety of elements might produce X rays, discovered by accident one day that uranium, a rare and little-studied material, could emit rays that would leave an image on a photographic plate without a separate light source. Examining this phenomenon further, Becquerel determined that some

element in uranium and, more powerfully, in the ore pitchblende produced rays that created images on photographic plates. It was this accident, on which Marie based her doctoral thesis research, that led to
her and her husband's own momentous discovery. On a request from
Becquerel, the couple began their work.

While it was known that pitchblende produced powerful rays of
what would later be called a radioactive nature, no real research had
yet been conducted into exactly what in the ore produced the actual rays. The Curies believed that a previously undiscovered element
must be present in the pitchblende. Their efforts were dedicated to
identifying that element. Pitchblende was mined in Bohemia, then a
part of the Austrian Empire. At their laboratory in Paris, the Curies
received from Austria a ton of the sandy material that they would use
to identify the central radioactive element they sought.

The Curies' initial research identified both uranium and the element
thorium as radioactive substances. Marie coined the term "radioactive" to describe the properties of the elements being studied. Hundreds of pounds of pitchblende (which contained both uranium and
thorium but emitted much more powerful rays than either element)
were expended through long months of effort by Pierre and Marie. The
husband-and-wife team finally isolated two new elements contained in
pitchblende, now distilled to salt form. The elements (identified in two
scientific papers presented to the Academy of Sciences in 1898) were
polonium (named for Marie's homeland) and radium. Further study
indicated that the trace element radium was far more powerful than
polonium. Based on her doctoral thesis outlining the success of her
experiments, Marie was awarded a doctorate in science from the Sorbonne in 1903. That same year, she, Pierre, and Henri Becquerel were
awarded the Nobel Prize in Physics for their work with radioactive
substances. The Curies were also awarded the prestigious Davy Medal
from the Royal Society in England.

The Curies' results took the turn-of-the-century scientific community by storm. Radium, then the most powerful radioactive substance
known to humankind, was considered one of the most important scientific discoveries in human history. The highly poisonous material
could burn unprotected hands, glowed of its own volition, emitted

massive amounts of heat, and was found, in properly controlled doses, to cure surface cancers on the skin. The Curies continued to experiment with the radium salts they had distilled from hundreds of pounds of pitchblende, refining the distillation process. Their work progressed until 1906, when Pierre was killed in a carriage accident on the streets of Paris. After his death, Marie was named as the first woman professor at the Sorbonne, replacing her husband in the physics chair he had occupied for several years prior to his death.

After Pierre's death, Marie Curie continued her research into radium and radioactivity. Her primary goal was to separate radium into its purest form. The distillation process she developed left the substance as part of a chloride-salt compound. In 1910, her experiments proved successful when she passed an electrical current through molten radium chloride, reducing the material to an alloy that, when further processed, became pure radium. For this breakthrough, she was awarded the 1911 Nobel Prize in Chemistry, making her the first person to receive two Nobel Prizes for scientific endeavors. The year 1913 saw her return to Warsaw to start the Radium Institute, a scientific facility dedicated to the education of men and women and to research into radioactivity. She returned to Paris near the end of the year.

The onset of hostilities in World War I saw Curie working feverishly both to prevent the radiological equipment and materials in France from falling into German hands and to make the medicinal benefits of radium available to soldiers in combat. She recruited and trained 150 young women as radiological-equipment operators to treat wounded soldiers. Her own daughter, Irène Joliot-Curie, was among the young women selected for this task. In addition to training young women, Curie also worked as an ambulance driver in France during the war. (Ambulances that were equipped with her X-ray machines were popularly known as "little Curies.")

In 1914, Curie was named director of the Sorbonne's Radium Institute, established in her and her late husband's honor. The institute made significant advancements in the use of X-ray technology and the medical applications of radiological technology during World War I. Her book on the scientific lessons learned from the war, *La Radiologie et la guerre*, was published in 1921.

By that year, Curie was one of the world's most beloved and renowned scientific figures. She traveled to the United States to receive a gift of an ounce of radium (one of only 150 ounces in existence at the time) from President Warren G. Harding for use in scientific research. She also made it one of her major priorities in the years after World War I to stockpile radium for scientific uses at the Paris Radium Institute. (A second visit to the United States in 1929 produced a meeting with President Herbert Hoover and funds to purchase another ounce of radium.) Marie's foresight in stockpiling radium helped advance medical applications of the substance and enabled other scientists, such as her daughter Irène and her son-in-law Frédéric Joliot, to conduct seminal research of their own. (The Joliot-Curies would go on to win the 1935 Nobel Prize in Chemistry for the creation of an artificial radioactive isotope.)

During her later years, Curie would continue to dedicate her life to the pursuit of knowledge in the area of radiological research. She lectured extensively, participated in numerous scientific assemblies and symposia, and continued to publish in her specialty. Her last book, *Radioactivité*, was published posthumously in 1935. She also participated in the formation of numerous scientific institutions that would help shape the nature of research for generations to come. In 1932, one of the most prominent of these facilities, the Curie Foundation, was started in her adopted home of Paris. It is a source of irony that Curie's life would be ended by the very substance she discovered. In May, 1934, she developed what was later identified as leukemia, brought on by prolonged exposure to radiation. On July 4, 1934, she died in Sancellemoz, France.

### Significance

The impact of the lives and work of Marie and Pierre Curie reached far beyond the specific scientific breakthroughs the two made. The Curies reshaped the landscape of modern physical understanding and built the foundation for many of the major discoveries that followed on their research. From having the curie, the basic unit of radiologic measure named for them, to participating in the creation of several pivotal scientific institutions, the Curies created a new scientific perspective as much as a new field of science through their work.

Beyond the science she developed, Marie showed the world caught up in the technological revolution of the late nineteenth and early twentieth centuries that women could play full and equal roles in the parade of progress that was reshaping the world. As the first female professor at the Sorbonne, she helped break the social barriers to scientific equality, influencing attitudes in later generations of scientists.

Pierre and Marie, as well as their daughter Irène and son-in-law Frédéric, would account for an astonishing three Nobel Prizes for scientific advancements. The Curies' story, as an example of pure dedication to the pursuit of knowledge, has also served as a model for children from their own time to modern times.

**Eric Christensen**

# Paul A. M. Dirac

## British physicist

*Dirac developed a general theory of quantum mechanics and formulated a relativistic wave equation that describes the behavior of an electron in an atom. The profound implication of Dirac's theory is that every particle has an antiparticle, which has been found to be correct. The theory is inherently a many-particle theory that provided the beginning of a theoretical framework for explaining how fundamental particles can be created and destroyed. The equation also explains electron spin and predicts the existence of antimatter.*

**Areas of Achievement**: Physics, mathematics

**Born**: August 8, 1902; Bristol, Gloucestershire, England

**Died**: October 20, 1984; Tallahassee, Florida

## Early Life

Paul A. M. Dirac (dih-RAK) was the second child born to Charles Dirac and Florence Holten Dirac. His father was born in Switzerland, while his mother was English. He had an older brother, Reginald, and a younger sister, Beatrice. Because of the strict discipline of his father, young Dirac was reared in an unhappy home, an experience that likely made him shy and withdrawn. His elementary school years were spent at Bishop Road Primary School, where he demonstrated his exceptional mathematical abilities at an early age. At the age of twelve, he attended Merchant Venturers Technical College, where his father taught French. Scientific subjects and modern languages were the primary emphasis of Dirac's secondary education.

In 1918, Dirac entered the University of Bristol. He completed a bachelor's degree in electrical engineering in 1921. Two years later, he earned a bachelor's degree in mathematics from the same institution. Shortly thereafter, he received a grant to participate in mathematical research at St. John's College in Cambridge, where he became interested in physics. During this time, Dirac's brother, Reginald, committed suicide, which left Dirac even more withdrawn from society.

Dirac's initial interest in physics focused on the theory of general relativity, but working at Cambridge under the direction his adviser, Ralph Fowler, Dirac pursued the study of quantum physics. In 1926, Dirac earned his doctorate in physics from Cambridge. His dissertation examined the development of a general theory of quantum mechanics. He published eleven papers on the subject prior to submitting his dissertation. After working with Niels Bohr in Copenhagen and with Max Born and other prominent physicists at the University of Gottingen in Germany, Dirac received a teaching appointment at Cambridge in 1927.

## Life's Work

At Cambridge, Dirac continued his research on quantum physics. In 1928, he reworked the basic wave equation of quantum mechanics (derived earlier by Erwin Schrödinger) so that Albert Einstein's special theory of relativity would be taken into account in the equation. This insight led Dirac to the derivation of the famous Dirac equation, a relativistic equation for the wave function that describes the motion of an electron in an atom.

The Dirac equation predicted the spin angular momentum of an electron and also the existence of the antielectron (later called positron), a particle that has the same properties as an electron, except for positive charge. Dirac published *Quantum Theory of the Electron* in 1928. In 1930, he published his landmark text on quantum mechanics, *Principles of Quantum Mechanics*, which has been revised and reprinted numerous times. A year later, he showed that the quantization of an electric charge could be explained by the existence of a magnetic monopole. Because the fundamental entity of magnetism is the dipole, a monopole has yet to be discovered.

In 1932, Dirac was appointed as the Lucasian Professor of Mathematics at Cambridge. For his work in helping to establish the foundations of quantum mechanics, Dirac shared the 1933 Nobel Prize in Physics with Schrödinger. Because of his quiet, private demeanor, Dirac initially thought about not accepting the award, but he accepted the prize, inviting his mother, but not his father, to the award ceremony in Stockholm, Sweden. During the 1934-1935 academic year, Dirac visited Princeton University, where he became good friends with physicist Eugene Wigner. In 1937, Dirac married Wigner's sister, Margit, and adopted her two children from a prior marriage. Dirac and Margit had two children of their own, Mary and Florence.

During 1937, Dirac became interested in cosmological research and devoted a great deal of time to that work. In 1939, he was awarded the Royal Medal of the Royal Society of London. While World War II was raging, he worked on the separation of uranium and nuclear weapons in Birmingham, England. For his many contributions to quantum mechanics, Dirac was awarded both the Max Planck Medal and the Copley Medal in 1952. He was known among his colleagues for his reserved, precise nature and his personal modesty. He enjoyed traveling and made several trips to the United States and the Soviet Union to present invited lectures.

Dirac was a member of numerous important academies and societies, including the Soviet Academy of Sciences, the Indian Academy of Sciences, the American Physical Society, the National Academy of Sciences, and the Royal Danish Academy of Sciences. He held numerous honorary degrees from many institutions. In 1969, he became the first recipient of the Oppenheimer Prize.

After retiring from Cambridge in 1969, Dirac accepted an appointment as a professor of physics at Florida State University (FSU) in 1971, so that he would be near his daughter, Mary. He was awarded the Order of Merit in 1973, a prestigious honor for intellectual achievement. He continued his physics research at FSU, concentrating on cosmological problems, until his death in 1984. The Paul A. M. Dirac Library at FSU was named in his memory. In his honor, the Paul Dirac Medal and Prize is awarded each year by the Institute of Physics, the professional society for physicists in the United Kingdom. The Abdus

Salam International Center for Theoretical Physics (ICTP) awards the ICTP Dirac Medal annually on Dirac's birthday. A plaque inscribed with his famous relativistic wave equation is maintained at Westminster Abbey in London.

## Significance

Dirac played a prominent role in the foundation and development of quantum mechanics. Using the work of Schrödinger and Werner Heisenberg, Dirac formulated the mathematical framework of quantum physics that associates measurable physical quantities with operators that act on the Hilbert space of vectors that represent a description of a physical system. In this formalism, Dirac introduced the very useful bra-ket notation and the Dirac delta function.

In 1927, Dirac developed a quantum theory of radiation, which was the beginning of quantum electrodynamics. That same year, he suggested the notion of second quantization, a way to transform the physical description of a single quantum particle into a mathematical formalism that describes the physics of a system of many such particles. In his book *Principles of Quantum Mechanics*, Dirac introduced many of the names and mathematical notations that are used in quantum mechanics, including fermion, boson, eigenfunction, physical observable, commutator, and the bra-ket notation.

Dirac's most famous discovery, the Dirac equation, describes the characteristics and behavior of the electron and predicts the existence of its antiparticle, the positron. Dirac's equation explained the origin of electron spin and the electron magnetic moment. Carl Anderson confirmed Dirac's prediction of antimatter when he discovered the positron in 1932. The profound implication of Dirac's theory is that every particle has an antiparticle, which has been found to be correct. The theory is inherently a many-particle theory that provided the beginning of a theoretical framework for explaining how fundamental particles can be created and destroyed.

Between 1932 and 1933, Dirac worked on a mathematical formalism that eventually led to the concept of renormalization used in quantum field theory. It provides ways of subtracting out unphysical, unwanted infinites that arise in quantum field theory so that the predictions for observable physical quantities are finite. Dirac and Enrico Fermi developed the Fermi-Dirac statistics, which is used to determine the distribution of electrons and other fermions among the available energy levels for a system composed of such particles.

**Alvin K. Benson**

# Albert Einstein

## German-born American physicist

*Einstein was the principal founder of modern theoretical physics, and his theory of relativity fundamentally changed the understanding of the physical world. His stature as a scientist, together with his strong humanitarian stance on major political and social issues, including nuclear energy and weaponry, made him one of the outstanding thinkers of the twentieth century.*

**Areas of Achievement**: Physics, astronomy, mathematics, peace advocacy

**Born**: March 14, 1879; Ulm, Württemberg, Germany

**Died**: April 18, 1955; Princeton, New Jersey

## Early Life

Albert Einstein (INE-stine) was born in Ulm, Germany, to moderately prosperous Jewish parents. His early childhood did nothing to suggest future greatness; he was late learning to speak, and his parents feared that he might be backward. He was apparently fascinated at the age of five by the mysterious workings of a pocket compass, and at the age of twelve he became enthralled by a book on Euclidean geometry. In his childhood, he also learned to play the violin and so acquired a love of music that was to last throughout his life.

In 1888, Einstein was sent to the Luitpold gymnasium in Munich, but he disliked the regimented and authoritarian atmosphere of the school. Even at this young age he seems to have exhibited the independence of mind, the ability to question basic assumptions and to trust in his own intuition, which were to lead him to his brilliant achievements. He left the gymnasium in 1895, without gaining a diploma. His father tried to send him to a technical school in Zurich, but he failed the entrance examination, in spite of high scores on mathematics and physics. The following year he was more successful, and in 1900, he received the diploma that qualified him to teach. To his disappointment, however, he failed to obtain a teaching position at the school.

In 1901, Einstein became a Swiss citizen. (He had renounced his German citizenship in 1896.) In 1902, after temporary positions at schools in Winterthur and Schaffhausen, he secured a post as a technical expert in the Swiss patent office in Berne, where he was to remain for seven years. In the following year, he married Mileva Maric, a friend from his student days in Zurich, and in 1904 the first of their two sons was born.

Einstein's first scientific paper had been published in 1901, and he had also submitted a Ph.D. thesis to the University of Zurich. While he was working quietly in the patent office, however, isolated from the mainstream of contemporary physics, there was little to suggest the achievements of 1905, which were to shake the scientific world to its core.

## Life's Work

In 1905, Einstein published three major papers, any one of which would have established his place in the history of science. The first, which was to bring him the Nobel Prize in Physics in 1921, explained the photoelectric effect and formed the basis for much of quantum mechanics. It also led to the development of television. The second concerned statistical mechanics and explained the phenomenon known as Brownian motion, the erratic movement of pollen grains when immersed in water. Einstein's calculations gave convincing evidence for the existence of atoms.

It was the third paper, however, containing the special theory of relativity, that was to revolutionize understandings of the nature of the physical world. The theory stated that the speed of light is the same for all observers, and is not dependent on the speed of the source of the light, or of the observer, and that the laws of nature (both the Newtonian laws of mechanics and Maxwell's equations for the electromagnetic field) remain the same for all uniformly moving systems.

This theory meant that the concept of absolute space and time had to be abandoned because it did not remain valid for speeds approaching those of light. Events that happen at the same time for one observer do not do so for another observer moving at high speed in respect of the first. Einstein also demonstrated that a moving clock would appear to run slow compared with an identical clock at rest with respect to the observer, and a measuring rod would vary in length according to the velocity of the frame of reference in which it was measured.

In another paper published in 1905, Einstein stated, by the famous equation $E = mc^2$, that mass and energy are equivalent. Each can be transferred into the other because mass is a form of concentrated energy. This equation suggested to others the possibility of the development of immensely powerful explosives.

Such was Einstein's achievement at the age of twenty-six. There had not been a year like it since Newton published his *Principia* in 1687. The scientific world quickly recognized him as a creative genius, and in 1909 he took up his first academic position, as associate professor of theoretical physics at the University of Zurich. After two more positions, one in Prague and the other in Zurich, he became a member of the Prussian Academy of Sciences and moved to Berlin in 1914.

In the meantime, Einstein had been working to extend the special theory of relativity to include new laws of gravitation, and the general theory of relativity was published in 1916. It was one of the greatest intellectual productions ever achieved by one person, and its picture of the universe as a four-dimensional space-time continuum lies at the foundation of all modern views of the universe. The theory stated that large masses produce a gravitational field around them, which results in the curvature of space-time. This gravitational field acts on objects and on light rays; starlight, for example, is deflected when passing through the gravitational field of the sun.

In 1919, the general theory received experimental confirmation from a team of British astronomers. Suddenly, the world awoke to the implications of Einstein's work, and he found himself internationally celebrated as the greatest scientist of the day. During the early 1920's, he traveled extensively in Europe, the Far East, and the Americas, hailed everywhere as genius, sage, and hero. With his untidy shock of hair formerly black, now graying rising from a high forehead, his deep brown eyes, and small moustache, he made a striking figure. It was not only his superior intellect that aroused public recognition and respect but also his simple good nature, nobility, and kindliness. However, Einstein, always modest, was genuinely astonished at the attention he received.

It was during his travels in the 1920's that the other great concerns of Einstein's life came to the fore. A man of deep humanitarian instincts, he did not isolate himself from the turbulent political events around him. During World War I, he had spoken out against militarism and nationalism. Now, as a famous person, he once more took up the cause of pacifism, expressing his opinion openly, caring nothing for popularity. Einstein's other lasting concern was the promotion of Zionism, and his tour of the United States in 1921 was undertaken in part to raise funds for Hebrew University. These activities made him a target for fierce abuse from the Nazis, and even outside Germany, his radical political views made him a controversial figure.

When Hitler came to power in 1933, Einstein was on his third visit to the United States, and he resolved not to return to Germany. After brief stays in Belgium and England, he left Europe for the last time, to become a professor at the Institute for Advanced Study at Princeton. He continued to lend his support to the cause of justice and freedom, helping Jewish refugees whenever he could, and he modified his former pacifism in the face of the threat of Nazi domination. In 1939, he was persuaded to write a letter to President Franklin D. Roosevelt, alerting him to the military potential of atomic energy. (Einstein played no part, however, in the research that led to the development of the atomic bomb, which merely verified the truth of his famous equation.) After the war, he remained tirelessly devoted to the cause of world peace, and proposed a world government in which all countries were to agree to forfeit part of their national sovereignty.

His later scientific career took two main directions. First, he was so deeply convinced of nature's fundamental simplicity that he labored unsuccessfully for thirty years in an attempt to construct a unified field theory. Second, he could not accept one of the fundamental results of quantum theory, that the interaction of subatomic particles could be predicted only in terms of probabilities. "God does not play dice with the world," he remarked. Many of his colleagues thought him stubborn, but nevertheless he remained a revered figure; his reputation as a genius who also possessed wisdom and saintliness never left him, neither in life nor in death.

## Significance

Einstein's scientific achievements place him alongside such figures as Copernicus, Galileo, and Newton, as one who vastly enlarged the scope of human knowledge about the physical universe. In this respect he is a universal figure and belongs to no country. It is perhaps appropriate, however, that Einstein, a German Jew to whom destiny had decreed a nomadic existence, eventually found a permanent home in the United States. His links with his adopted country he became a U.S. citizen in 1941 are profound. He was the most illustrious of the hundreds of intellectuals who fled from Europe before World War II, and his presence at the newly formed Institute for Advanced Study, which marked a new period of development for American research and education, played a key role in attracting other eminent scholars.

Einstein had always viewed the United States as a bulwark of democracy and individual freedom, and in the debates that divided the country in the postwar decade particularly debates on the Cold War and the use of nuclear energy his was a consistent voice for sanity and decency in human affairs. He spoke out for freedom of thought and speech in the McCarthy era, when he feared that the United States was betraying its own ideals, and he continued to urge scientists to consider the social responsibilities of their work in the atomic age. In his final year, he and a group of leading scientists signed a statement, known as the Russell-Einstein Manifesto, warning about the terrible consequences of nuclear war. This led to the Pugwash Conference on science and world affairs in 1957, in which for the first time scientists from East and West met to discuss nuclear arms. A series of influential conferences followed, and the Pugwash movement has continued its activities ever since.

**Bryan Aubrey**

# Michael Faraday

## British physicist

*Considered by many to have been the greatest British phys-
icist of the nineteenth century, Faraday made discoveries
in electromagnetism that were fundamental to the development of field physics. His inventions of the dynamo and
electric motor provided the basis for modern electrical industry.*

**Areas of achievement**: Physics, chemistry

**Born**: September 22, 1791; Newington (now in
London), Surrey, England

**Died**: August 25, 1867; Hampton Court, Surrey,
England

### Early Life

Michael Faraday (FAHR-ah-day) was the third of four children born
to James Faraday, a Yorkshire blacksmith, and Margaret Hastwell, the
daughter of Yorkshire farmers. Both were of Irish descent. Shortly before his birth, the family moved to Newington, near London, in search
of better opportunities. James Faraday's health deteriorated, limiting
his ability to work, and the family had only the bare necessities for
survival. Young Faraday's education consisted of the rudiments of
reading, writing, and arithmetic. The family belonged to the small religious sect of Sandemanians, which emphasized the Bible as the sole
and sufficient guide for each individual, and Faraday was a devoted,
lifelong member.

In 1804, Faraday was an errand boy for George Riebau, a London
bookseller and bookbinder. His seven-year apprenticeship produced
an extraordinary manual dexterity, a skill characteristic of his experimental researches. He also read omnivorously, from *The Arabian*

*Nights' Entertainments* to the *Encyclopaedia Britannica*. The latter's article on electricity awakened him to a new world, as did Jane Marcet's *Conversations on Chemistry* (1806), a book that converted him in his teenage years into a passionate student of science. With his apprenticeship nearing an end in 1812, it was not likely that Faraday would be anything but a bookbinder. A customer's gift of tickets to a series of lectures by Humphry Davy at the Royal Institution changed his life. Davy was a scientist of international stature and a brilliant lecturer, largely responsible for the success of the Royal Institution, a center both for research and for the dissemination of science to a general audience.

Faraday, enthralled by the lectures, desperately wanted to become a scientist. When Davy became temporarily blinded in a laboratory explosion, a customer at Riebau's bookshop recommended Faraday to him as secretary for a few days because of Faraday's fine penmanship. Faraday subsequently bound in the bookshop his neatly written lecture notes with his own illustrations and sent the volume to Davy asking for a job. Davy had nothing for him at the time. Suddenly, in 1813, however, Davy fired his laboratory assistant for brawling, and the twenty-one-year-old Faraday became his assistant.

In that same year, Davy married a rich widow and set out on a grand tour of Europe, including visits to the major scientific centers to meet the most distinguished Continental scientists. Faraday went along to assist Davy in his research. The tour was a remarkable experience; the young man had never been more than a few miles from London. His letters home were full of amazement over meeting renowned scientists during the eighteen-month tour. On his return to England, the Royal Institution appointed Faraday superintendent of apparatus. Now in his early twenties, he possessed a robust intelligence, considerable scientific knowledge, and the good fortune to be at the Royal Institution.

All Faraday's contemporaries described him as kind, gentle, and simple in manner. Serenity and calm marked his life and countenance; no scientist has been referred to more as humble or saintly. These attributes stemmed from his Sandemanian faith, with its stress on love and community. He had an unquestioning belief in God as creator and sustainer of the universe and saw himself as merely the instrument by which the divine truths of nature were exposed. His faith and his science meshed completely.

Otherworldly, Faraday had a contempt for moneymaking and trade, and he rejected all honors that raised him above others. He refused both knighthood and the presidency of the Royal Society. In 1821, he married a fellow Sandemanian, Sarah Barnard. The marriage was childless but most happy. She lavished her maternal feelings on the nieces who lived with them and on her husband. United by a deep, enduring love, secure in their faith, the tone of the household (they lived in rooms provided in the Royal Institution) was one of gaiety, and domestic life was completely satisfactory.

## Life's Work

Faraday was a late bloomer with no important discovery until he was more than thirty. He lacked familiarity with mathematics, the language of physics, and remained outside the mathematical tradition of universities and of Continental physics. From 1815 to 1820, he earned a modest reputation as an analytical chemist, publishing several papers on subjects suggested by Davy. These were the years of his scientific apprenticeship.

In 1820, the Danish physicist Hans Christian Oersted discovered the magnetic effect of the electric current. This discovery of electromagnetism caused a sensation and provoked both an explosion of research and much confusion. In 1821, the editor of a journal asked Faraday to review the experiments and interpretations and present a coherent account of electromagnetism. Faraday's genius now became evident, for he demonstrated that there were no attractions or repulsions involved in the phenomenon; instead, a force in the conductor made a magnetic needle move around it in a circle. He also devised an instrument to illustrate the process, producing the first conversion of electrical into mechanical energy. He had discovered electromagnetic rotation, and as a by-product, he had invented the electric motor.

Faraday did not follow up this major discovery with anything comparable until 1831, although his chemical researches continued to be fruitful, notably the 1825 discovery of benzene, which he isolated from an oil that separated from illuminating gas. He also conducted a lengthy project for the Royal Society on the improvement of optical glass used in lenses. It ended with no apparent useful results, but he did prepare a heavy lead borosilicate glass that later proved indispensable to his electromagnetic work.

In 1825, the Royal Institution promoted Faraday to director of the laboratory. Faraday instituted the Friday Evening Discourses, which soon became one of the most famous series of lectures on the progress of science, serving to educate the English upper class in science and to influence those in government and education. In 1826, he began the Christmas Courses of Lectures for Juvenile Audiences, which further extended the appeal of the institution. His lectures, based on a careful study of oratory, were full of grace and earnestness, and exercised a magic on hearers. He was at his best with children: a sense of drama and wonder unfolded, and they reacted with enthusiasm to the marvels of his experiments. Two of his courses for juveniles were published as *The Chemical History of a Candle* (1861) and *The Various Forces of Nature* (1860). They have remained in print as classics of scientific literature.

In 1831, Faraday made his most famous discovery, reversing Oersted's experiment by converting magnetism into electricity. He used the Royal Institution's thick iron ring as an electromagnet, winding insulated wire on one side with a secondary winding on the other side. With a battery linked to one winding and a galvanometer to the other, he closed the battery circuit and the galvanometer needle moved. He had induced another electric current through the medium of the iron ring's expanding magnetic force. He called his discovery electromagnetic induction and elaborated a conception of curved magnetic lines of force to account for the phenomenon.

Over the next several weeks, Faraday devised variations and extensions of the phenomenon, the most famous one being the invention of the dynamo. He converted mechanical motion into electricity by turning a copper disc between the poles of a horseshoe magnet, thereby producing continuous flowing electricity. From this discovery came the whole of the electric-power industry. Faraday realized that he had a possible source of cheap electricity, but he was too immersed in discovery to pursue the practical application.

In 1833, Faraday made his most monumental contribution to chemistry. A study of the relationship between electricity and chemical action disclosed the two laws of electrochemistry. He then devised a beautiful, elegant theory of electrochemical decomposition that involved no poles, no action at a distance, no central forces. Faraday's theory, totally at odds with the thinking of his contemporaries, demanded a new language for electrochemistry. In 1834, in collaboration with the classical scholar William Whewell, he invented the vocabulary of electrode, anode, cathode, anion, cation, electrolysis, and electrolyte, the word *electrode* meaning not a pole or terminal but only the path taken by electricity. Faraday's stupendous labors of the 1830's were too much for him, however, and in 1838, he suffered a serious mental breakdown. So bad was his condition that he could not work for five years.

In 1845, William Thomson (Lord Kelvin) suggested to Faraday some experiments with polarized light that might reveal a relation between light and electricity. This stimulated Faraday into intense experimentation. He had no success until he tried a stronger force, an electromagnet, and passed a polarized light beam through the magnetic field. At first unsuccessful, he remembered his heavy borosilicate glass from the 1820's. Placing it between the poles of the magnet, he sent the light beam through the glass and the plane of polarization rotated; he had discovered the effect of magnetic force on light (magneto-optical rotation).

The fact that the magnetic force acted through the medium of glass suggested to Faraday a study of how substances react in a magnetic field. This study revealed the class of diamagnetics. Faraday listed more than fifty substances that reacted to magnets not by aligning themselves along the lines of magnetic force (paramagnetics) but by setting themselves across the lines of force, a finding that attracted more attention from scientists than any of his other discoveries.

During the 1850's, Faraday's theorizing led to the idea that a conductor or magnet causes stresses in its surroundings, a force field. The energy of action lay in the medium, not in the conductor or magnet. He came to envision the universe as crisscrossed by a network of lines of force, and he suggested that they could vibrate and thereby transmit the transverse waves of which light consists. (The notion of the electromagnetic theory of light first appeared in an 1846 Royal Institution lecture.) His speculations had no place for Newtonian central forces acting in straight lines between bodies, or for any kind of polarity. All were banished for a field theory in which magnets and conductors were habitations of bundles of lines of force that were continuous curves in, through, and around bodies.

Faraday's mental faculties gradually deteriorated after 1855. Concern for his health reached Prince Albert; at his request, Queen Victoria in 1858 placed a home near Hampton Court at Faraday's disposal for the rest of his life. There, he sank into senility until his death in 1867. Like his life, his funeral was simple and private.

## Significance

Michael Faraday was an unusual scientist. He never knew the language of mathematics. To compensate, he had an intuitive sense of how things must be, and he organized his thoughts in visual, pictorial terms. He imagined lines of force stretching and curving through the space near magnets and conductors. In this way, he mastered the phenomena. His vision of reality was incomprehensible to a scientific world preoccupied with the Newtonian model. Only when James Clerk Maxwell showed how Faraday's ideas could be treated rigorously and mathematically did the lines-of-force conception in the guise of field equations become an integral part of modern physics.

Faraday coupled his inventive thinking with an unmatched experimental ability. His ingenuity disclosed a host of fundamental physical phenomena. One of those phenomena, his seemingly humble discovery of the dynamo, became the symbol of the new age of electricity, with its incalculable effects on society and daily life.

**Albert B. Costa**

# Enrico Fermi

## Italian-born American physicist

*Fermi's experiments utilizing neutron bombardment led to the production of the first controlled chain reaction, critical to the development of the atomic bomb by the United States.*

**Areas of Achievement**: Physics, mathematics

**Born**: September 29, 1901; Rome, Italy

**Died**: November 28, 1954; Chicago, Illinois

## Early Life

Enrico Fermi (ayn-REE-koh FEHR-mee) was born to Alberto Fermi, an administrator in the Italian railroad system, and Ida de Gattis, an elementary school teacher. Enrico learned to read and write at an early age, probably instructed by his older brother and sister. At six years of age, Enrico entered public school and soon displayed a talent for mathematics. By the age of ten, the young boy, by this time exhibiting a remarkable memory, entered a school that emphasized Latin, Greek, French, history, mathematics, natural history, and physics, courses that would prepare him for entrance into the university. Enrico led his class in scholarship and, in his spare time, studied science and built electrical motors and toys.

In 1918, after receiving his diploma from the *liceo* (high school) and winning on a competitive examination, Fermi was admitted as a fellow in the Scuola Normale Superiore, a college of the University of Pisa. The young scholar earned his doctorate, graduating magna cum laude in 1922 at age twenty-one, and obtained a postdoctoral fellowship to study physics under Max Born in Göttingen in 1923. During the academic year of 1923–1924, Fermi held a temporary position at the University of Rome as an instructor of mathematics. Then, in

September of 1924, Fermi began a three-month fellowship at the University of Leiden. He later accepted a nontenured position as *incaricato* (instructor) at the University of Florence, teaching mechanics and mathematics.

Fermi's physical appearance at this time, according to his future wife, was not as impressive as his mental abilities. His rounded shoulders, short legs he was five feet six inches tall thin lips, a neck thrust forward when he walked, and a dark complexion were in sharp contrast to his gray-blue, close-set, cheerful eyes. Vital and energetic, he particularly enjoyed skiing and hiking.

### Life's Work

In November, 1926, Fermi won the *concorso* (competition) for a new chair in theoretical physics at the University of Rome, a result, in part, of the publication of his paper "On the Quantization of the Perfect Monoatomic Gas." This paper calculated the behavior properties ("Fermi's statistics") of an ideal gas composed of particles of half integral spin (electrons, protons, and so on).

In 1928, Fermi published a book on modern physics for upper-level university students, *Introduzione alla fisical atomica*. During this period as professor at the university, he successfully recruited talented students to study physics with him and other faculty members. On July 19, 1928, the young physicist married Laura Capon in Rome. Two children were born of the union, Nella in 1931 and Giulio in 1936. In 1929, Fermi was appointed as the youngest member of the Royal Academy of Italy with a government salary as part of the honor, and in the summer of 1930, he was invited to teach theoretical physics at the University of Michigan, Ann Arbor, lecturing on the quantum theory of radiation.

In 1933, Fermi wrote a famous paper on the explanation of beta decay. After the announcement of Frédéric Joliot and Irène Joliot-Curie in Paris that alpha particle bombardment of aluminum produced artificial radioactivity, the Italian physicist experimented with neutrons as the bombarding source. He was able to produce artificial radioactivity in fluorine in March, 1934, using a radon-plus-beryllium source of neutrons. Seven months later, in October, 1934, Fermi discovered a principle of nuclear physics that was to have far-reaching effects on the future of science. Placing a piece of paraffin in front of a neutron source, he observed increased radioactivity in the silver target. He surmised that the paraffin showed the neutron "bullets" and increased the neutron-proton collision cross section. This allowed the silver nuclei to capture the slow neutron, eject a proton, and become temporarily radioactive. Fermi and his coexperimenters, Bruno Pontecorvo, Edoardo Amaldi, Franco Rasetti, and Emilio Segre, also noted that water produced almost the same slowing-down effect as paraffin. The theory was proposed that neutrons lose energy in repeated collisions with hydrogen nuclei. Anticipating possible commercial applications, the scientists took out an Italian patent on this neutron-bombardment process of producing radioactive substances in October, 1935.

Fermi's reputation in scientific circles in the United States grew, and in the summer of 1936, he was invited to give a course on thermodynamics at Columbia University in New York. On previous teaching trips in 1933 and 1935, Fermi was impressed by the freedom and kindness of the American people. This appreciation had prompted him earlier to consider moving to the United States to escape the dictatorship of the regime of Benito Mussolini in Italy. Although Fermi had little interest in politics, he did understand that free and open scientific investigation was more desirable than that practiced under the dictates of an oppressive government. In 1938, Italy, influenced by the anti-Semitism sweeping Germany at that time, passed laws against persons of Jewish ancestry. Since Fermi's wife was Jewish, he decided to leave Italy. He accepted a teaching position at Columbia and obtained an immigration visa in November, 1938. The next month, Fermi was awarded the Nobel Prize in Physics for his work with slow neutrons, and on December 24, 1938, the Fermi family left Europe for the United States. (Fermi became a U.S. citizen in July, 1944.)

Events in nuclear physics were developing rapidly in late 1938. On December 22, 1938, Otto Hahn and Fritz Strassmann in Germany published the results of their experiments on the neutron bombardment of uranium. They had discovered radioactive barium among the products of that bombardment. In January, 1939, Otto Frisch and Lise Meitner theorized that the presence of radioactive barium indicated the fission (splitting) of the uranium nucleus into two nuclei of approximately equal size, barium and krypton. Fermi then surmised that if enough excess neutrons were released in the fission process, and if enough uranium atoms were present, a chain reaction, with the release of enormous amounts of energy, might result.

In early 1939 at Columbia, Fermi exchanged experimental information with two Hungarian emigrant scientists, Leo Szilard and Edward Teller. In August, 1939, Albert Einstein, the most famous physicist in the world, informed President Franklin D. Roosevelt by letter that the work of Fermi and Szilard demonstrated the possibility that a powerful bomb might be constructed, utilizing the nuclear chain-reaction principle. Roosevelt formed a Committee on Uranium to keep him apprised of the progress of the experimentation.

In the spring of 1940, Fermi and others discovered the use of graphite as a moderator in slowing down neutrons. This principle would be vital in future chain-reaction experiments involving a nuclear reactor ("pile"). Fermi also observed that lumping natural uranium would permit the start of a chain reaction without the need of isotope separation a process that, given the technology of the time, seemed to be an almost impossible task.

By the summer of 1940, scientists at the University of California identified the product of the fission of natural uranium 238 as neptunium 239. Neptunium decays and produces plutonium 239, which fissions when bombarded with slow neutrons. If a chain reaction utilizing uranium 238 could be sustained, then enough fissionable plutonium 239 could be obtained for use in the manufacture of an atomic bomb.

During 1941, Fermi experimented with a small atomic pile at Columbia University. He showed that a self-sustaining nuclear chain reaction could be achieved if the proper amount of uranium was placed in a graphite pile. By the spring of 1942, Arthur Holly Compton, professor of physics at the University of Chicago, had been placed in charge of all work pertaining to the chain reaction. Compton brought physicists to Chicago to coordinate the experimental effort under the code name "Metallurgical Laboratory." Fermi supervised the construction of a large pile under the stands of the university's football stadium beginning in October, 1942. On December 2, 1942, cadmium control rods were slowly withdrawn; a twenty-eight-minute, self-sustaining chain reaction occurred, producing one-half watt of energy. This was the first nuclear reactor to produce energy that would eventually be used for peaceful purposes. Several months later, a second pile (one hundred kilowatts) was built at the Argonne Laboratory outside Chicago. This success led to the construction, with Fermi's consultation, of the large water-cooled, natural uranium-fueled, plutonium-production reactor at Hanford, Washington, in 1944.

Near Los Alamos, New Mexico, in an isolated mesa area, a new laboratory was constructed for the purpose of building an atomic bomb. The best minds in nuclear physics were brought to Los Alamos or consulted regarding the bomb construction. The first experiments began in July, 1943, under the direction of J. Robert Oppenheimer of the University of California, Berkeley. Fermi was still conducting work in Chicago and traveling to Oak Ridge, Tennessee, the site of a plant that produced uranium 235, a fissionable isotope that would be used in the manufacture of yet another bomb. He came to Los Alamos in August, 1944, and was appointed as an associate director of the laboratory by Oppenheimer. During this period, Edward Teller was working on the hydrogen bomb under Fermi's direction.

By July, 1945, enough reactor-produced plutonium had been delivered to Los Alamos to allow for the production of the first atomic bomb. The testing of this weapon occurred on July 16, 1945, at Alamogordo, New Mexico. Fermi was responsible for measuring the energy levels produced by the explosion and for collecting sand and soil samples to be analyzed for radioactivity.

The first atomic bomb was dropped on Hiroshima, Japan, on August 6, 1945, and the war ended shortly thereafter. In January, 1946, Fermi returned to Chicago as Distinguished-Service Professor of Physics at the Institute of Nuclear Studies, and in that year, he and four other scientists were awarded the Congressional Medal of Merit by President Harry S. Truman for their work in developing the bomb. For the next eight years, Fermi continued his experiments in high-energy physics. During this period, he also served on the General Advisory Committee of the Atomic Energy Commission and on other advisory bodies in need of his scientific knowledge and counsel. In April, 1954, Fermi testified at the security risk hearing of Oppenheimer, whose loyalty to the United States had been questioned because he had opposed the program for developing the hydrogen bomb. (Fermi himself, as well as other influential scientists, including Einstein, had opposed the development of this thermonuclear weapon on ethical grounds.) At the hearing, Fermi pointed out Oppenheimer's service to the United States and suggested that there should be no question of Oppenheimer's loyalty.

By the summer of 1954, Fermi's health was deteriorating. An undiagnosed illness that later proved to be cancer drained his energy. He died in Chicago on November 28, 1954, at the age of fifty-three.

## Significance

Fermi epitomized the popular image of the scientist, the diligent experimenter in lab coat surrounded by the apparatus of discovery, yet he was much more. He was a talented mathematician and theorist who was able, if necessary, to build analytical instruments with his own hands. His keen mind searched for simple alternatives to problems that others thought impossible. He was an immigrant to the United States, imbued with a spirit of gratitude and loyalty to his new country. He was a devoted husband and father. Although he helped develop the atomic bomb as a weapon of war, a weapon that he hoped would not be used on a civilian population, his name is primarily associated with the controlled chain reaction that is the basis of the peaceful use of atomic energy in nuclear power plants. Fermi the immigrant, like so many immigrants to America before him, brought to this nation a new knowledge, a new prestige, influence in world politics, and the basis of a new power. Fermi has entrusted future generations with the responsible use of that power.

**Charles A. Dranguet, Jr.**

# James Clerk Maxwell

## Scottish physicist

*Both a theoretical and an experimental physicist as well as a notable mathematician, Maxwell founded modern field theory and statistical mechanics, mathematically describing interactions of electrical and magnetic fields that produce radiant energy, thus confirming the existence of electromagnetic waves that move at light speed. He also elaborated theories of the mechanics and kinetics of gases and a theory of Saturn's rings.*

**Area of Achievement**: Physics

**Born**: June 13, 1831; Edinburgh, Scotland

**Died**: November 5, 1879; Cambridge, England

### Early Life

An only child, James Clerk Maxwell was reared by devout Episcopalian parents who, in a generally impoverished Scotland, enjoyed the comforts of Middlebie, a modest landed estate, and other small properties. Though rarely practiced, his father's profession was law. Between his parents and grandparents, James was a descendant of middle-level government officials, landed developers of small mines and manufactories, and acquaintances of such famous Scotsmen as Sir Walter Scott and the great geologist John Hutton, although none of Maxwell's immediate kin displayed unusual drive or distinction.

Maxwell's boyhood was that of a happy, unusually observant, and well-loved child to whom were imparted clear religious and moral precepts that prevailed throughout his life. His health was sometimes delicate, and between his fourteenth and sixteenth years he learned that he was both nearsighted and afflicted with a persistent ear infection. Regardless of whether these infirmities were contributory, he manifested a shyness and reserve, a superficial impression of dullness, though not unfriendliness, that remained permanent characteristics.

In his earliest years, Maxwell had no formal education. Clearly, despite his affection for the outdoors, his mother introduced him to John Milton's works, as well as other classics, and he became a catholic, voracious reader prior to his entrance into the prestigious Edinburgh Academy in 1841. There, for six years, if at first unenthusiastically, he pursued a classical curriculum. Equally important, Maxwell's father introduced him into the meetings of the Edinburgh Society of Arts and the Royal Society, where he met D. R. Hay, a decorative painter interested in explaining beauty in form and color according to mathematical principles.

Stimulated by his own prior interest in conic forms, Maxwell was encouraged to pursue such inquiries seriously, one result being his receiving of the academy's Mathematical Medal (he also took first prize in English verse) in 1846. A second consequence stemmed from his introduction to Dr. James D. Forbes, of Edinburgh University, subsequently a lifelong friend, for Forbes sent the young Maxwell's ". . . Description of Oval Curves, and Those Having a Plurality of Foci" to be included in the *Proceedings of the Edinburgh Royal Society* the same year. At the age of fifteen, Maxwell, in sum, was already recognized by Forbes and other mentors as an original, proficient, and penetrating mind, confirmation of this coming with publication by the Royal Society of two additional papers in 1849 and 1850, one dealing with the theory of rolling curves, the other with the equilibrium of elastic solids. These achievements came while Maxwell attended the University of Edinburgh, steeping himself in natural philosophy (a Scottish intellectual specialty of great logical rigor), mathematics, chemistry, and mental philosophy.

## Life's Work

Precocious, already credited with natural genius, Maxwell entered Cambridge University (first Peterhouse but soon Trinity College) in 1850. There, he swiftly came under the direction of William Thomson, popularly known for helping lay the Atlantic cable but academically to gain fame as Lord Kelvin, expert on the viscosity of gases and collaborator with James Joule in experiments on properties of air, heat, and electricity and in thermodynamics. Maxwell's superb tutor, William Hopkins, simultaneously brought discipline and order to Maxwell's incredible range of knowledge. Graduated in 1854, Maxwell shortly was made a Fellow of Trinity College and was authorized to lecture. By 1856, however, he accepted a professorship in natural philosophy at Marischal College, whose reorganization as the University of Aberdeen in 1860 caused him to move to King's College, London, still professing natural philosophy. He took with him to London the daughter of Marischal's principal, Katherine Dewar, whom he had married in 1858. They were to have no children.

Maxwell remained at King's until 1865. From students' perspectives, he was a poor instructor; his voice, mirroring his shyness, was husky and monotonal; his explanations were pitched beyond their grasp, particularly when he was lecturing workingmen. In addition, both his wife's health and his own seemed precarious. Upon arrival at King's, he had been infected with smallpox, and in 1865 a riding accident resulted in erysipelas, which seriously drained him. Perhaps persuaded by these circumstances, he retired to the family farm at Glenair until 1871, when Cambridge University offered him a new chair in experimental physics. Because the university's chancellor had presented funds to it for a modern physical laboratory, subsequently the world-famous Cavendish Laboratory, Maxwell devoted himself to designing and equipping it, alternating between spending academic terms there and summers at Glenair. He completed this task in June, 1874.

Contrary to superficial appearances, Maxwell's theoretical and experimental work proceeded steadily from his entrance into Trinity College until his death. Before completion of the Cavendish, he had converted his London home into an extensive laboratory, and the results of his investigations were both continuous and impressive. They basically fell into seven seemingly disparate but actually related areas: experiments in color vision and optics, which later had important consequences in photography; studies in elastic solids; explorations in pure geometry; mechanics; Saturn's rings; and electromagnetism and electricity, which began with Michael Faraday's lines of force and eventuated in Maxwell's theory of the electromagnetic field, in the electromagnetic theory of light, and, among other electrical investigations, in establishing standards for the measurement of resistance for the British Association.

Maxwell published in each of these seven areas, but he is undoubtedly best known for his works on electricity and magnetism. His papers "Faraday's Lines of Force" and "Physical Lines of Force," presented between 1855 and 1861, were seminal studies. Originally confessing little direct knowledge of the field in which Michael Faraday worked, Maxwell nevertheless sought to demonstrate mathematically that electric and magnetic behavior was not intrinsic to magnetic bodies or to conductors, that rather, this behavior was a result of vaster changes in the distribution of energy throughout the ether, albeit by unknown means.

In a 1864 paper titled "On a Dynamical Theory of the Electromagnetic Field," Maxwell further demonstrated that electromagnetic forces moved in waves and that the velocity of these waves in any medium was the same as the velocity of light, thus paving the way for an electromagnetic theory of light. In connection with his many published papers, Maxwell also wrote a textbook, *Theory of Heat* (1871), a study in dynamics, *Matter and Motion* (1876), and gathered and edited the *Electrical Researches of Henry Cavendish* (1879). His own *An Elemental Treatise on Electricity* was published posthumously, in 1881.

While Maxwell recovered from his erysipelas and remained in good health until 1877, though in his prime, he fell ill again with painful dyspepsia. For nearly two years, Maxwell treated himself and kept silent on his illness. By 1879, when he acknowledged it to physicians, his disease was diagnosed as terminal, and he died at Cambridge on November 5, 1879.

### Significance

James Clerk Maxwell is probably the only nineteenth century physicist whose reputation became greater in the twentieth century than it had been in his own century. Rarely have such capacities for inventiveness, exposition, experiment, and mathematical descriptiveness been brought to bear by one man in the physical sciences. He gave new direction and substantiation to Faraday's work and effected a bridge to the investigations of Heinrich Hertz, who did in fact measure the velocity of electromagnetic waves, confirm that these waves indeed behaved precisely like light, and showed therefore that light and electromagnetic waves were one and the same. Practical evidence of the importance of this work is manifest in modern electronics, in radio, television, and radar.

Maxwell was the effective founder of field theory and of statistical mechanics, with enormous implications for theoretical and tabletop physics, not only for questions that he clarified but also for those that he raised. Similarly, his curiosity about the rings of Saturn led him productively into study of the kinetic theory of gases, adding to the work of John Herapath, Rudolf Clausius, and Joule by treating the velocities of molecules statistically. His conclusions about the nature of light made physical optics a branch of electricity, providing a basis for the study of X rays and ultraviolet light. Even in metallurgy, he was credited with the invention of an automatic control system. In short, his inquiries ranged from the macrocosm to the microcosm and were ultimately knit together and described mathematically in the tradition of Newton.

**Clifton K. Yearley**

# Sir Isaac Newton

## English physicist

*Newton's theory of gravitation and laws of mechanics described, for the first time, a natural world governed by immutable physical laws. In addition to creating a conceptual framework that underlay the practice of science until the twentieth century, Newton's understanding of the world in terms of natural laws profoundly affected the history of ideas and the practice of philosophy in the modern era.*

**Areas of achievement**: Physics, science and technology

**Born**: December 25, 1642 (new style, January 4, 1643); Woolsthorpe Manor, near Colsterworth, Lincolnshire, England

**Died**: March 20, 1727 (new style, March 31, 1727); London, England

### Early Life

Sir Isaac Newton was born on Christmas Day, 1642, to a farmer and his wife, at Woolsthorpe Manor, just south of Grantham in Lincolnshire. His father died shortly before Newton's birth, and when his mother remarried three years later and moved away to live with her new husband, Newton remained at Woolsthorpe to be reared by his grandparents.

Newton attended the grammar school in Grantham, and he demonstrated scientific aptitude at an early age, when he began to construct mechanical toys and models. Aside from a brief period when his mother tried to persuade him to follow in his father's footsteps and become a farmer, his education continued (it is said that Newton tended to read books rather than watch sheep, with disastrous results). He was accepted as an undergraduate at Trinity College, Cambridge, in 1661.

Although his mother provided a small allowance, Newton had to wait on tables at college to help finance his studies. Even at that time, his fellow students remarked that he was silent and withdrawn, and indeed, Newton throughout his life was something of a recluse, shunning society. He never married, and some historians believe that Newton had homosexual leanings. Whatever the truth of this speculation, it is certain that he preferred work, study, experimentation, and observation to social activity, sometimes to the detriment of his own health. After retreating to Grantham for a short time while Cambridge was threatened by plague, Newton returned to the university as a don in 1667 with an established reputation for mathematical brilliance.

## Life's Work

It was not long at all before Newton proved his reputation for genius to be well deserved. Shortly after his graduation, Newton developed the differential calculus, a mathematical method for calculating rates of change (such as acceleration) that had long evaded other scholars. As a result, in 1669, he was offered the Lucasian Chair of Mathematics at Cambridge, a position he held until 1701.

Newton's second major contribution of this period was in the field of optics. His experiments with light had led him to build a reflecting telescope, the first one of its kind that actually worked. After further refinements, he presented the device to the Royal Society, where he was asked to present a paper on his theory of light and colors. Shortly afterward, he was made a fellow of this august body, which contained all the prominent intellectuals of the day.

Newton's paper offered new insights into the nature of color. While experimenting with prisms, Newton had discovered that white light is a mixture of all the colors of the rainbow and that the prism separates white light into its component parts. Newton's theory was controversial, provoking strong feelings at the Royal Society and initiating a lengthy dispute with Robert Hooke concerning the nature of light. Hooke criticized Newton with such vehemence that Newton presented no more theories on the nature of light until 1704, after Hooke's death.

For a scientist such as Newton, the seventeenth century was an interesting period during which to work. Scientific thought was still dominated by the Aristotelian worldview, which had held sway for more than two thousand years, but cracks in that outlook were beginning to appear. Galileo had shown that the planets traveled around the Sun, which was positioned at the center of the universe, while Johannes Kepler had observed that this motion was regular and elliptical in nature. The task confronting scientists, in keeping with the aim of explaining the universe mathematically from first principles, was to find some logical reason for this phenomenon.

Newton, among others, believed that there had to be a set of universal rules governing motion, equally applicable to planetary and earthbound activity. His researches finally led him to a mathematical proof that all motion is regulated by a law of attraction. Specifically, he proved that the force of attraction between two bodies of constant mass varies as the inverse of the square of the distance between those bodies (that is, $F_A = k/D^2$, where $F_A$ is the force of attraction between the bodies, D is the distance between them, and k is a constant). From this beginning, he was able to explain why planets travel in ellipses around the Sun, why Earth's tides move as they do, and why tennis balls, for example, follow the trajectories that they do. The inverse square formula also led Newton toward a notion of gravity that neatly tied his mathematics together. When Newton published this work, it led to another major confrontation with Hooke, who claimed that he had reached the proof of the inverse square law before Newton; the argument between the two was lengthy and acrimonious.

In 1684, Edmond Halley, then a young astronomer, went to Cambridge to visit Newton, who was reputed to be doing work similar to Halley's. Halley found that Newton claimed that he had proved the inverse square law but had temporarily mislaid it. (Throughout his life, Newton worked on scraps of paper, keeping everything from first drafts to final copies, so this assertion has the ring of truth to it.) Halley was astounded: Here was a man who claimed to have solved the problem that was bothering many leading scientists of the day, and he had not yet made it public. When Halley returned, Newton had found the proof, and Halley persuaded him to publish his nine-page demonstration of the law. Still, Halley was not satisfied. Realizing that Newton had more to offer the world, he prodded him into publishing a book of his theories. The result was the famous *Philosophiae Naturalis Principia Mathematica* (1687; *The Mathematical Principles of Natural Philosophy*, 1729, best known as the *Principia*), which was published at Halley's expense. A year later, a second and third volume of the work reached the public.

The *Principia* was a highly technical and mathematical work that many of Newton's contemporaries had difficulty following, but its effect on the scientific community was profound. In it, Newton outlined his three laws of motion. The first states that every body continues in a state of rest or motion until it is acted on by a force. The second law states that the acceleration of a body is proportional to the force applied to it and inversely proportional to its mass. The third law, perhaps the most widely quoted, states that for every action there is an equal and opposite reaction. From these three fundamental laws, Newton went on to construct his theory of gravity—a force that acts at a distance between two or more bodies, causing an attraction between them that is in inverse proportion to the distance between them.

Newton's theories were a major challenge to the dominant worldview, constructing the world, as they did, purely from mechanics. His theories seemed revolutionary and initiated a great debate, which continued for the better part of a century after the *Principia* was published. When they were eventually accepted as a useful description of nature, Newtonian science formed the basis of modern thinking until the twentieth century, when Albert Einstein's theories turned the world upside down again.

Writing the *Principia* dominated Newton's life to such an extent that he became completely obsessed with the project, often forgetting to eat or even to sleep while he continued working. Despite his reclusive tendencies, however, in 1687 Newton entered the public arena. Cambridge University and King James II, a Catholic, were in the midst of a battle over religion. The university had refused to grant a degree to a Benedictine monk, and the officials of the university, including Newton, were summoned to appear before the infamous Judge George Jeffreys to argue their case. Shortly afterward, Newton was elected the member of Parliament for Cambridge. Newton's entrance into politics was less than world-shattering, though; it is said that he spoke only once during his term of office, and that was to ask an usher to open a window.

In 1693, Newton suffered a mental breakdown about which little is known, and he withdrew into his previous solitary state. Two years later, he returned to public office when he was asked to take over the wardenship of the mint. There was to be a major reissue of coinage because of the increasingly pressing problem of clipped gold and silver coins. New coins needed to be minted with milled edges, and several prominent scientists were pressed into service to aid in the process. Newton discovered a hitherto unrecognized penchant for administration and proved himself a highly able bureaucrat, being promoted to master of the Mint in 1699. In 1701, he was reelected to Parliament and continued in the public eye for the remainder of his life.

Until his death in 1727, honors were heaped upon Newton, as befitted the most prominent scientist of his generation. In 1703, he was elected president of the Royal Society and was annually reelected to that post for the next twenty-five years. He moved to London and became more sociable but nevertheless earned a reputation for being cantankerous and ill-tempered. In 1704, Newton published *Opticks*, a tract about the theories of light that he had earlier expounded to the Royal Society. It was more accessible than the *Principia* and gained a wider audience. A year later, he was knighted by Queen Anne. Meanwhile, the *Principia* was proving to be a best-seller, as everyone wanted to read the theories that were pushing back the frontiers of contemporary science, and it went through second and third editions during Newton's lifetime.

Newton's work in the last years of his life was mainly religious, apart from another acrimonious dispute with the German philosopher Gottfried Wilhelm Leibniz over who had first invented the differential calculus. Newton spent hours attempting to understand the messages hidden in the Book of Revelation, seeing this task as simply another aspect of the search for truth as revealed in God's works, both written and created. Thus, in the end, Newton proved himself to be a medieval thinker, despite his work laying the foundations of modern scientific thought.

## Significance

Newton made an outstanding contribution to the modernization of the Western scientific worldview. He followed in the footsteps of Nicolaus Copernicus, Galileo, Kepler, and others in asserting that the heavens and earth were a part of one solar system (not separated as they are in Aristotelian philosophy), with the Sun at the center. Newton further developed and refined the method of observation and experiment that had already established itself in the seventeenth century, by carefully checking and rechecking his work and by creating experimental verifications of his various theories. Most important, he demonstrated that a comprehensive mechanical description of the world that explained matter and motion in terms of mathematics was actually possible. With the *Principia*, Newton effectively sounded the death knell of the old description of the universe and laid the basis for a modern approach. His was perhaps the greatest individual contribution to a rich and innovative period of scientific development.

**Sally Hibbin**

# Max Planck

## German physicist

*Planck's discovery in 1900 that light consists of infinitesimal "quanta" and his articulation of the quantum theory replaced classical physics with modern quantum physics. This work not only resulted in Planck's receiving the Nobel Prize in Physics in 1918 but also became a major enabling factor in the work of many other Nobel laureates.*

**Area of Achievement**: Physics

**Born**: April 23, 1858; Kiel, Schleswig (now in Germany)

**Died**: October 4, 1947; Göttingen, West Germany (now in Germany)

### Early Life

Born into an intellectual family, Max Planck (plahngk) spent most of his early life in Munich, where the family moved in the spring of 1867, when he was nine. Planck's father, Johann Julius Wilhelm von Planck, was a professor of civil law at the university in Kiel, whose second wife, Emma Patzig, was Max's mother. Max's forebears included many lawyers and clergymen, a fact that helps explain Planck's lifelong respect for the law and interest in religion.

In May, 1867, Planck was enrolled in Munich's Königliche Maximilian-Gymnasium, a classical gymnasium (secondary school), where he came under the tutelage of Hermann Müller, a mathematician who took an interest in the youth and taught him astronomy and mechanics as well as mathematics. It was from Müller that Planck first learned the principle of the conservation of energy that underlay much of his future work in thermodynamics and quantum theory.

On completion of the gymnasium in 1874, Planck was at a personal crossroads. He was gifted in music and humanities as well as in mathematics and the sciences. He concluded ultimately that his music talents were insufficient to justify his continuing in that field. When he entered the University of Munich in October, 1874, he concentrated on mathematics. At Munich, however, his interest in physics grew, although his mathematics professors tried to dissuade him, arguing that nothing new remained to be discovered in the field.

Planck became ill as he was completing his first year at Munich and missed two years of school. In the winter term of 1877-1878, when he was well enough to resume his studies, he entered the University of Berlin, where he decided to study theoretical physics because of the order and logic that discipline demanded. Planck yearned to study the nature of the universe. Theoretical physics offered him his most sensible foothold for achieving that goal.

In Berlin, Planck studied with Hermann von Helmholtz, Gustav Kirchhoff, and Rudolf Clausius. Only Clausius was a gifted teacher, although the other two were able physicists. Planck learned much on his own through reading. Although his doctoral dissertation, on the second law of thermodynamics, was undistinguished, he was graduated summa cum laude in 1879. He taught mathematics and physics briefly at his former secondary school in Munich and in 1880 was appointed a privatdocent at the University of Munich. At that time, theoretical physics was viewed as an unpromising field, so his future seemed less than bright.

In 1885, Planck became an associate professor of physics at the University of Kiel, where he remained until 1888, when he was appointed assistant professor and director of the Institute for Theoretical Physics, to replace Kirchhoff, who had died. He rose to professor in 1892 and remained at Berlin until his retirement in 1926.

### Life's Work

Planck's early work in the laws of thermodynamics and his early interest in the principle of the conservation of energy figured largely in his research from his early teaching days at Kiel through his first decades at the University of Berlin. Although he had been reared on classical physics and was a conservative at heart, Planck began to realize that the laws of classical physics deviated greatly from results obtained in experimental physics. He found the greatest disparities not in the field of optics but in that of thermodynamics. The problems stemmed from the measurement of radiant energy in the frequency spectrum of black bodies.

Kirchhoff deduced that radiant energy is independent of the nature of its radiating substance, reasoning that black bodies that absorb all frequencies of light should therefore radiate all frequencies of light. Energy at that time was considered infinitely divisible, a theory that led to many anomalies and seeming contradictions in physics. The problem arose because the lower-frequency range has a smaller number of frequencies than the higher-frequency range.

Important physicists working on this problem reached conflicting conclusions. Wilhelm Wien devised an equation that explained the emissions at high frequencies but not at low frequencies. John William Strutt, third baron of Rayleigh, devised an equation that worked for low frequencies but not for high frequencies. Work in the field was at an impasse when Planck devised a classically simple equation that explained the distribution of radiation over the full range of frequencies, basing his equation on the daring supposition that energy is not an indivisible flow but is composed of tiny particles, or "quanta," for the Latin word meaning "How many?" Incidental to this discovery was his discovery of a means of measuring the absolute weight of molecules and atoms, in itself a major breakthrough.

Planck's theory showed that the energy of various frequencies of light from violet to red contain different energies, a quantum of violet containing twice the energy of a quantum of red and requiring twice the energy to radiate from a blackbody, making such radiation improbable. So pristine and uncluttered was Planck's theory that he himself was suspicious of it.

Other scientists, however, began to realize its validity, and soon Albert Einstein based much of his work on photoelectric effect, which classical physics could not explain, on quantum theory. Planck embraced Einstein's theory of relativity eagerly because of its absolutism and because of its presentation of the velocity of light.

Now firmly established at the University of Berlin, Planck was instrumental in bringing Einstein to the Berliner Academie in 1914 as a professor without teaching obligations and as director of the embryonic Kaiser Wilhelm Institute for Physics, which Planck himself eventually headed. Planck also nominated Einstein for the Nobel Prize in Physics in 1921, a year in which the award was withheld. In 1922, however, partly at Planck's urging, Einstein received the 1921 prize a year late.

Max Born, Theodor von Kármán, and Peter Joseph Wilhelm Debye began to study the problem of the dependence of specific heat on temperature from the standpoint of quantum theory and soon articulated a law that made it possible to ascertain the variation in specific heat with temperature from the elastic constants in any substance. The field of quantum mechanics became the most important field of physics in the first half of the twentieth century, followed closely by the field of quantum electrodynamics, both developments that evolved from Planck's original insights and from his expression of the ratio between the size of a quantum and its frequency by the symbol $h$, which expresses a universal quantity.

Planck, a balding, bespectacled man with heavy brows, a dark mustache, and grayish eyes, spent nearly four decades at Berlin, teaching extensively and carrying heavy administrative responsibilities. He apparently was a splendid, well-organized teacher, who was clear in his presentations and interested in students. His life during these years was not easy. His wife died in 1909, and, in the next decade his son was killed in World War I and his two daughters died in childbirth.

With Adolf Hitler's rise to power, Planck decided that he had to remain in Germany, although he deplored what was happening. His respect for the law was deeply ingrained, and he felt duty-bound as a citizen to live within the laws but to work from within to change them. He intervened unsuccessfully for Jewish friends and colleagues who were being sent to death camps.

As a Nobel laureate of enormous prestige, Planck scheduled an interview with Hitler and tried to dissuade him from the genocide that was overwhelming Nazi Germany. Hitler, on learning why Planck had come to see him, began a diatribe that lasted for hours; Planck's intervention did not deter Hitler from his disastrous course. Before the end of the war, Planck, in his eighties, had lost his home and all of his papers to a bombing raid, had once been trapped for several hours in a collapsed air-raid shelter, and, worst of all, had suffered the execution by the Nazis of his son Erwin, a secretary of state before Hitler's ascension, who had been accused of plotting to assassinate Hitler.

## Significance

Planck lived another twenty-one years after his retirement. These were troubled years in Germany. The search for the meaning of the universe and for the nature of existence that had led him into physics, where he hoped he would discover absolutes to help answer his questions, persisted in his later years. He wrote on general subjects, developing some of his earlier lectures and essays into fuller works.

In 1930, Planck became president of the Kaiser Wilhelm Society of Berlin, which was renamed the Max Planck Society in his honor. In his final postwar years, he again became president of the society, agreeing as he approached his ninetieth year to assume the post until a permanent president could be found.

Five volumes of Planck's work in theoretical physics were published in English under the title *Introduction to Theoretical Physics* (1932-1933). His highly philosophical *Physikalische gesetzlichkeit im lichte neuer forschung* (1926) and *Das Weltbild der neuen Physik* (1929; combined in *The Universe in the Light of Modern Physics*, 1931) were released and showed a search for absolutes in a broadly religious context, although Planck sought a prime cause more than an anthropomorphic god.

His general works *Where Is Science Going?* (1932) and *Wege zur Physikalischen Erkenntnis* (1933; *The Philosophy of Physics*, 1936) were combined with *The Universe in the Light of Modern Physics* and published in English under the title *The New Science* (1959). Planck's autobiography *Wissenschaftliche Selbstbiographie* (1948; *Scientific Autobiography and Other Papers*, 1949) was published posthumously.

When the Allies came into Germany in May, 1945, Planck, who, with his second wife, had fled to Magdeburg to live with friends after the destruction of their home near Berlin, was again homeless. The area was overrun with Allied soldiers, and Planck had no place to live. American soldiers rescued him and had him sent to a hospital in Göttingen, the city in which he lived for the two and a half years remaining to him. He continued his professional activities, giving his last public lecture on pseudoproblems in 1946.

**R. Baird Shuman**

# Ernest Rutherford

### New Zealand physicist

*One of the pioneers of the atomic age, Rutherford investigated the nature of the atom and of radioactivity, laying the experimental basis of the new atomic physics.*

**Areas of achievement**: Physics, science

**Born**: August 30, 1871; Spring Grove (later known as Brightwater), near Nelson, New Zealand

**Died**: October 19, 1937; Cambridge, England

## Early Life

Ernest Rutherford was the fourth child in a family of twelve, and although his father was an engineer, he spent his early life on a farm. He had to travel some distance to school but quickly established himself as a student of ability. In 1886, he won a scholarship to Nelson College (with a result that was a record for the times). Three years later, he won a university scholarship to Canterbury College, Christchurch, from which he was graduated in 1894 with a double first in physical sciences and mathematics.

While at Christchurch, Rutherford met Mary Newton, the only recorded romantic attachment of his life. He had, in fact, a reputation for being somewhat diffident with women. He began what was to become a six-year courtship, feeling secure enough to marry only in 1900, when he had taken a professorship in Canada, after several years in Europe.

In 1894, Rutherford won the 1851 Exhibition Scholarship, an award established to finance young graduates from the British Empire to study in England. He had already published an audacious paper challenging the leading scientists of the day, titled "Magnetisation of Iron by High Frequency Discharges." He designed and constructed the apparatus himself, no mean achievement for a twenty-three-year-old, isolated from the scientific community in New Zealand. In 1895, therefore, it was no surprise that he was courted by the Cavendish Laboratory in Cambridge; he went to study there at one of the most exciting periods in its history, working alongside such eminent scientists as James Clerk Maxwell, J. J. Thomson, and John William Rayleigh. It was the year of the discovery of X rays, a scientific development that was to point the way to Rutherford's future work.

Rutherford was at first something of an outsider in the rarefied atmosphere of Cambridge University. Not only was he not a graduate of that university, but also his New Zealand country upbringing made him appear both brash and gauche. Moreover, unlike most of his colleagues, he had no independent means, and despite various small scholarships, he had to pay his way by taking on work as a tutor. He was a large, ruddy-faced man with a directness of manner that people who did not know him well considered uncouth. It was, however, this very characteristic that later, when he ran his own laboratory, inspired loyalty among his colleagues.

His first research was on the long-distance transmission of radio waves. He had, in fact, arrived from New Zealand with a radio-wave detector that he himself had constructed. Indeed, one of Rutherford's chief characteristics was his ability to devise experimental apparatus to test the leading theories of the day. Although his research into radio waves was important Guglielmo Marconi was working along similar lines at the time and a few years later developed the wireless Rutherford's thoughts began to turn in other directions. J. J. Thomson, who was something of a mentor to the young Rutherford, was working on X rays, and Rutherford elected to join him.

### Life's Work

One of Rutherford's early experimental successes was to detect and measure the ionization caused by the passage of electricity through gases. He was working with Thomson's theory of ionization; the experiment was important in that it showed Rutherford's ability to translate theory into a measurable reality a talent that he used throughout his life. In 1898, Rutherford was offered the second MacDonald Professorship of Physics at McGill University, Montreal, Canada.

Although his work with Thomson at the Cavendish was over, he continued his research into what would later become known as radioactivity. A year later, he published a paper based on his work at the Cavendish, which showed that there were two distinct types of radiation: alpha and beta rays. He was working toward demonstrating that the energy of alpha and beta rays comes from within the atom. He discovered a curious "emanation" from thorium that is, radioactivity which decreased geometrically with time. His own discovery, however, was overshadowed by the Curies' discovery, in the same period, of radium and the half-life of its emanation. Historically, therefore, the discovery of radioactivity is ascribed to the Curies. Rutherford's greatest achievement in this field occurred when he joined forces with Frederick Soddy an unusual combination, since Soddy was a chemist. Together they analyzed the emanation, realizing that it arose from the transformation of one element into another. That was, at the time, an outrageous suggestion, with its overtones of alchemy, but within a brief period the scientific community came to accept that radioactivity, as it was now called, was a new kind of phenomenon. In 1908, Rutherford was awarded the Nobel Prize in Chemistry for his work on radioactivity, a unique achievement for a physicist and an indication of the unorthodox combination of disciplines whereby Rutherford had achieved his goal.

In 1907, Rutherford left Canada. He had begun to feel isolated from the main centers of scientific research and turned down offers from Yale University and the Smithsonian Institution to return to Great Britain and take up the Chair of Physics at Manchester University. At a time when European scientists such as Werner Heisenberg and Albert Einstein were leading the world in scientific theory, Rutherford established a school at Manchester that set out to unite experimental and theoretical talent. Among the world-famous scientists who studied under him were Lawrence Bragg (later professor of chemistry at Manchester), Harry Moseley (who was killed during World War I), Hans Geiger (who later invented the Geiger counter), and Niels Bohr, perhaps the most brilliant scientist of the twentieth century. Under Rutherford's leadership, Manchester became one of the major scientific centers of the world.

Rutherford was now a leading figure in the scientific community, corresponding with scientists around the world, advising appointments in Europe, Australia, and North America, and playing a major part in the Royal Society. His most important international contribution at this time was to take part in the convention to establish an international radium standard. This was a very delicate task, particularly given Marie Curie's prior claims in the field. It demanded a high level of diplomacy from the normally blunt Rutherford, and the convention was eventually successful. His international standing was honored at home when he was knighted in 1914.

Rutherford now pointed his research in a slightly different direction. He had already, as has been noted, established the nature of radioactivity; now he used radioactivity to determine the nature of the atom itself. The currently accepted model was that of Rutherford's old teacher, Thomson, who visualized the atom as a kind of "plum pudding," with the atomic particles suspended haphazardly within its confines. This model, however, was at odds with Max Planck's quantum theory and could not explain the discrete units of energy emitted by atoms. Through continual correspondence and experimentation, Rutherford, Planck, and, later, Bohr (after he had left Manchester) gradually developed a new model, the one that has become generally accepted. If Planck and Bohr separately pushed out the theoretical boundaries of atomic physics, then Rutherford and his colleagues (first at Manchester, later at the Cavendish Laboratory) supplied the experimental verifications.

Bohr's suggested model of the atom is the basis of the contemporary theory. It pictures negatively charged electrons orbiting around a central nucleus composed of positively charged protons and uncharged neutrons. It was Rutherford's carefully planned experiments that provided the first indications that Bohr's theory was correct, and later, in 1917, when Rutherford smashed the atom and the predicted particles were emitted, Bohr's theory was demonstrated beyond doubt. In disproving Thomson's version, Rutherford had finally overtaken his teachers' teachings.

During World War I, Rutherford, like many of his colleagues, was called into service on behalf of the federal government. Rutherford's particular contribution was in the field of antisubmarine warfare, and he worked on numerous committees and with the navy to discover ways of detecting submarines underwater, concentrating on sonic methods. Once again, Rutherford's talents for diplomacy were called on to construct a team of scientists and naval officers to work together under difficult and sometimes hostile circumstances. His experiences during that time convinced him that scientific research and government policy had to be linked both financially and in terms of priorities, and, after the war, he set out to establish more formal contacts between the two arenas. While Rutherford spent much time and energy on his war work, it was eventually superseded by his atomic work, particularly during the excitement of 1917, when it became clear that his ambition of smashing the atom was nearing fruition. As always, for Rutherford, science came first and anything else was pushed into the background when important new discoveries were on the horizon.

In 1919, Thomson resigned as head of the Cavendish Laboratory, and inevitably Rutherford was asked to take his place. As the Professor of the Cavendish, Rutherford was responsible for its postwar regeneration, and he reorganized and expanded it to become the world's leading center of experimental physics. Further, he developed it into a modern institution, in which teaching and research were planned to fulfill national needs.

Rutherford continued to play a part on the political stage in the 1930's. When many Jewish scientists were having to flee from Nazi Germany, Rutherford became a champion of the refugees, helping them in both practical and political matters; in particular, he enabled the Nobel Prize winner Max Born to work at the Cavendish until he succeeded in obtaining the chair at Edinburgh. Rutherford believed that science was an international pursuit without geographical boundaries, and these beliefs were put to a severe test when Pyotr Leonidovich Kapitsa, a brilliant student from the Soviet Union, came to study at the Cavendish in 1921. Thirteen years later, he was detained in Leningrad and not allowed to return to the West. Rutherford worked tirelessly to find a means whereby Kapitsa could continue his work in either Great Britain or the Soviet Union and finally succeeded in transferring the bulk of Kapitsa's experimental apparatus from Cambridge to Moscow perhaps the most difficult diplomatic task of his life.

Under Rutherford's direction at the Cavendish, Sir James Chadwick organized the finances and the research policy of the laboratory, Sir John Cockcroft and E. T. S. Walton built their famous accelerator, and the Cavendish led the world in experimentation on the atom and its composition. Rutherford achieved personal honor when he was elected president of the Royal Society, a post he held from 1925 to 1930. He died of intestinal paralysis in 1937.

## Significance

During Rutherford's scientific career, the theories of physics had changed beyond recognition: The discovery of radioactivity and the smashing of the atom were two of the most important milestones of the atomic age, and Rutherford had played a major part in both developments. Under his guidance, the Cavendish Laboratory had reached the limits of its experimental possibilities. The new era demanded larger and larger equipment, and only international institutes that could pool the resources of several governments could carry on the research. The unanswered questions were about the structure of the nucleus, and higher energies requiring larger accelerators were needed to probe its secrets. Rutherford had been one of the key experimentalists of the atomic age, but the new directions in physics were to take it into the nuclear age. Rutherford died before the major practical applications of his work, nuclear power and nuclear weaponry, were developed. Many believe that, while he would have welcomed the former, he would have striven to limit the potential of the latter. He was without doubt one of the founders of the nuclear age. He was buried in Westminster Abbey as Lord Rutherford of Nelson.

**Sally Hibbin**

# Bibliography

"A Greene Universe." *Scientific American* 21 Apr. 2000: 36. Print.

"And the Nobels go to ... Illinois." *Chicago Tribune* 8 Oct. 2003: 1. Print.

"Benjamin Franklin Medal Awarded to Dr. Sumio Iijima, Director of the Research Center for Advanced Carbon Materials, AIST." aist.go.jp. AIST, Feb. 2002. Web. 7 Aug. 2012.

"Catching the Light of a Baby Supernova." *Space Daily*. SpaceDaily.com, 26 May 2008. Web. 9 Aug. 2012.

"Dr. Aprille J. Ericsson." *Visiting Researcher Profile*. National Center for Earth and Space Science Education, Sept. 2007. Web. 9 Aug. 2012.

"Dr. Janna Levin Discusses Her Idea of a Finite Universe." *Talk of the Nation/Science Friday*. Natl. Public Radio, 12 July 2002. Web. 8 Aug. 2012.

"Dr. Katharine Hayhoe Talks About California's Climate Change." *Talk of the Nation*. NPR. 27 Aug. 2004. Radio.

"Female Frontiers." *NASA Quest*. NASA. Web. 5 Apr. 2012.

"Femtosecond Comb Technique Vastly Simplifies Optical Frequency Measurements." *Physics Today* June 2000: 19. Print.

"Fermilab's Helen Edwards Receives Prestigious 2003 Robert R. Wilson Prize from the American Physical Society." *Fermilab Press Pass*. U.S. Department of Energy, 22 Oct. 2002. Web. 9 Aug. 2012.

"Giving Mars Back Its Heartbeat. Great Terraforming Debate: Part I." *Space Daily*. SpaceDaily.com, 16 June 2004. Web. 8 Aug. 2012.

"Gray and Muddy Thinking About Global Warming." realclimate.org. Web. 5 Apr. 2012.

"Helen Edwards is Associate Head of Booster Section." *The Village Crier* 2.11 (19 Mar. 1970). Print.

"Hugh Herr." *Biomechatronics Group*. Massachusetts Institute of Technology, n.d. Web. 26 Jan. 2012.

"Jonathan Lethem and Janna Levin." *Seed Magazine*. Seed Media Group, 5 Mar. 2007. Web. 8 Aug. 2012.

"Katharine Hayhoe." *The Secret Life of Scientists and Engineers* (*Nova* web video series). PBS, 2011. Web. 19 Dec. 2011.

"Lemelson-MIT Awards Two $30,000 Student Prizes for Innovation." *PR Newswire* 9 Feb. 2000.

"MacArthur Foundation Names 31 Recipients of 1988 Awards." *New York Times* 19 July 1988: A23. Print.

"Physicist Shares Nobel Prize for Medicine." *PhysicsWorld.com*. Institute of Physics, 6 Oct. 2003. Web. 9 Aug. 2012.

"Quiet Man Making a Big Bang." *Scotland on Sunday* 14 Sept. 2008: 19. Print.

"Red Dragon Mission: A Discussion with Chris McKay." *The Mars Quarterly* 4.1 (Summer 2012): 4–5. Print.

"Stefan Rahmstorf." *PIK-Potsdam.de*. Potsdam Institute for Climate Impact Research. Web. 8 Aug. 2012.

"Sumio Iijima." *BusinessWeek*. Bloomberg L.P., 7 July 2002. Web. 7 Aug. 2012.

"Sumio Iijima." *Nano Tsunami*. Voyle.com. Web. 7 Aug. 2012.

"Superfluids: An Interview with Dr. Deborah Jin." *ESI Special Topics*. Research Services Group of Thomson Scientific, July 2005. Web. 9 Aug. 2012.

"The Double Amputee Who Designs Better Limbs." *Fresh Air*. Natl. Public Radio, 10 Aug. 2011. Web. 26 Jan. 2012.

Abrikosov, A. A. "My Years with Landau." *Physics Today* Jan. 1973: 56–60. Print.

Adelson, Eric. "Best Foot Forward." *Boston*. Metrocorp, Mar. 2009. Web. 27 Jan. 2012.

Astor, Dave "Four Decades for Forecasting Firm: Newspaper List Growing as Accu-Weather Marks Anniversary." *Editor & Publisher* 23 June 2003: 21. Print.

Battersby, Stephen. "Opinion Interview: Janna Levin, Theoretical Cosmologist." *New Scientist* 6 Apr. 2002: 40. Print.

Booth, William. "Why Not Implant Life on Mars?" *Washington Post* 17 Dec. 1990. Print.

Broadway, Bill. "When Theology Meets Cosmology; Physicist Wins Prize for Work on Science and Spirituality." *Washington Post* 11 March 1995: B7. Print.

Brooks, Michael. "Interview: The Final Frontier." *New Scientist* 18 June 2005: 88. Print.

Brumfiel, Geoff. "Did Design Flaws Doom the LHC?" *Nature*. Nature Publishing Group, 23 Feb. 2010. Web. 17 Jan. 2012.

Butler, Sharon. "University of Chicago Particle Physicist Maria Spiropulu is…" *Chicago Tribune* 11 July 2002: T1. Print.

Colquhoun, Kate. "Scents and Acute Sensibilities." *Daily Telegraph* 17 June 2006: 6. Print.

Cookson, Clive. "Domestic Science." *Financial Times*. Financial Times, 23 Sep. 2011. Web. 17 Jan. 2011.

Cousteau, Fabien. "Biography." *Fabien Cousteau: The Adventure of Discovery*. Fabien Cousteau, n.d. Web. 5 Apr. 2012.

Crace, John. "Lisa Randall: Warped View of the Universe." *The Guardian* 21 June 2005: 20. Print.

Cutter, Kimberly. "Who Is the Nouveau Cousteau?" *Men's Journal* (2008): n. pag. Print.

Daley, Beth. "'Genius' Has Its $500,000 MacArthur Awards Three Hub Residents Among 23 Recipients." *Boston Globe* 28 Sept. 2004: B1. Print.

Davidson, Keay. "Heated debate on formation of solar system/Theorist believes 9 planets created out of chaos, not calm." *San Francisco Chronicle* 17 June 2002: A6. Print.

Dembart, Lee. "Book Review: Universe That Knows What It's Doing." *Los Angeles Times* 18 Mar. 1988: 16. Print.

Dizikes, Peter. "Across the Universe." *Boston Globe* 4 Sept. 2005: E1. Print.

Downing, Bob. "Scorching Summers a Definite Possibility." *Ohio.com*. Akron Beacon Journal, 29 July 2009. Web. 19 Dec. 2011.

Dreifus, Claudia. "Searching for Extraterrestrial Life." *New York Times* 21 July 2009: D2. Print.

---. "With Findings on Storms, Centrist Recasts Warming Debate." *New York Times* 10 Jan. 2006: F2. Print.

Ecenbarger, William. "Divining the Skies." *Chicago Tribune* 11 Jan. 1987: V1. Print.

Finn, Robin. "PUBLIC LIVES; Heir to an Undersea World, Swimming With Sharks." *New York Times* 31 July 2002: 2. Print.

Flatow, Ira. "WMAP Project and How It's Answering Questions About the Universe." *Talk of the Nation/Science Friday*. National Public Radio, 21 Feb.2003. Web. 9 Aug. 2012.

Folger, Tim. "Cosmologist David Spergel-Decoder of the Cosmos." *Discover Magazine* May 2003. Print.

Gewin, Virginia. "Movers: Rolf-Dieter Heuer, Director-General, CERN, Geneva, Switzerland." *Nature* 451.7178 (2008): 602. Web. 17 Jan. 2012.

Goodell, Jeff. "The Secretary of Saving the Planet." *Rolling Stone* 25 June 2009: 58+87. Print.

Greenberg, Andy. "A Step beyond Human." *Forbes*. Forbes.com, 25 Nov. 2009. Web. 25 Jan. 2012.

Gugliotta, Guy. "D.C. Scientist Shares Nobel in Physics; Three Are Honored for Space X-Ray, Neutrino Discoveries." *Washington Post* 9 Oct. 2002: A2. Print.

Hadhazy, Adam. "Deborah Jin Keeps It Cool With Quantum Mechanics." *Scientific American*. Nature America, 4 May 2009. Web. 9 Aug. 2012.

Hansch, Theodor. "Edible Lasers and Other Delights of the 1970s." *Optics and Photonics News* Feb. 2005: 14. Print.

Hara, Yoshiko "Blue Laser is Born of Bright Spirit." *Electronic Engineering Times* 30 Oct. 1997: 262. Print.

---. "Laser, LED Developer Dreams in Color." *Electronic Engineering Times* 24 Apr. 2000: 189. Print.

Hayhoe, Katharine. "Thermometers Don't Lie: An Interview with Author

Herrera, Stephan. "Weather Wise." *Forbes* 14 June 1999: 90. Print.

Hill, Stephen. "Sniff 'n' Shake." *New Scientist* 2115 (3 Jan. 1998): 3434. Print.

Johnson, George. "New Generation of Physicists Sustains a Permanent Revolution." *New York Times* 20 June 2000: F1. Print.

Johnstone, Bob. "True Boo-Roo." *Wired Magazine* 3.03 (Mar. 1995). Print.

Judson, Horace Feeland. "No Nobel Prize for Whining." *New York Times* 20 Oct. 2003. Print.

Kahn, Jennifer. "The Extreme Sport of Origami." *Discover Magazine* 29 July 2006. Print.

Karagianis, Liz. "Inventing Solutions." *Spectrum*. Massachusetts Institute of Technology, Fall 2000. Web. 8 Aug. 2012.

Kasey, Pam. "Atmospheric Scientist Offers View of a Changed West Virginia." *State Journal* [West Virginia] 9 Feb. 2007: 6. Print.

Kennedy, Louise. "Unfolding Origami's Secrets." *Boston Globe* 11 Nov. 2004. Print.

Kennedy, Pagan. "Necessity Is the Mother of Invention." *New York Times Magazine* 30 Nov. 2003: 86. Print.

Kintisch, Eli. "Nobelist Gets Energy Portfolio, Raising Hopes and Expectations." *Science* 19 Dec. 2008: 1774–75. Print.

Kluger, Jeffrey. "The 2006 TIME 100: Kerry Emanuel." *Time Magazine* 8 May 2006: 92. Print.

Kotulak, Ronald. "Seriously Weird Science." *Chicago Tribune* 11 January 2004. Print.

Kunzemann, Thilo. "Taking on Climate Change Myths and Skeptics." *Allianz Knowledge*. Allianz, 9 June 2009. Web. 8 Aug. 2012.

Lanchester, John. "Odor Decoder." *New York Times Sunday Book Review* 3 Dec. 2006: 32. Print.

Lawrence, Felicity. "Pater Mansfield, the Doctor Under Scrutiny." *The Guardian*. Guardian News and Media Limited, 7 Aug. 2001. Web. 9 Aug. 2012.

Lemonick, Michael D. "Astrophysics: Mr. Universe." *Time Magazine*. Time, 20 Aug. 2001. Web. 9 Aug. 2012.

Loll, Anna-Cathrin. "Lord of the Particles." *Asia Pacific Times*. Asia Pacific Times, Sep. 2009. Web. 18 Jan. 2012.

Mansfield, Peter. "Autobiography." *NobelPrize.org*. Nobel Foundation, Oct. 2003. Web. 9 Aug. 2012.

Marder, Jenny. "Hunt for Higgs Continues; Scientists Work to Separate the 'Signal from the Noise.'" *The Rundown*. MacNeil/Lehrer Productions, 13 Dec. 2011. Web. 18 Jan. 2012.

May, Mike. "Interview: Steven Chu." *American Scientist* Jan./Feb. 1998: 22. Print.

McCarthy, Alice. "Hugh Herr: Back on Top." *Science Careers*. Amer. Assn. for the Advancement of Science, 20 June 2003. Web. 29 Jan. 2012.

McFarling, Usha Lee. "3 Awarded Nobel Prize in Physics." *Los Angeles Times* 9 Oct. 2002: 110. Print.

Miller, Gerri. "Cousteau, the Next Generation." *Arts & Culture*. MNN Holdings, LLC, 21 Apr. 2009. Web. 5 Apr. 2012.

Minicucci, Daniela. "Spree of Deadly Storms Points to 'Global Weirding.'" *Global Toronto*. Shaw Media, 26 May 2011. Web. 19 Dec. 2011.

Minkel, JR. "Astronomers Witness Supernova's First Moments." *Scientific American*. Nature America, 21 May 2008. Web. 9 Aug. 2012.

Mirsky, Steve. "Paul Davies: Physics Could Help Fight Cancer." *Scientific American*. Nature America, 13 Apr. 2011. Web. 9 Aug. 2012.

Morgan, James. "Back to the Future." (Glasgow, Scotland) *Herald* 7 Apr. 2008: 13. Print.

Moss, Frank. The Sorcerers and Their Apprentices: How the Digital Magicians of the MIT Media Lab Are Creating the Innovative Technologies That Will Transform Our Lives. New York: Crown, 2011. Print.

Mufson, Steven. "Concern for Climate Change Defines Energy Dept. Nominee." *Washington Post*. The Washington Post Company, 12 Dec. 2008. Web. 5 Apr. 2012.

Murray, Charles J. "Bionic Engineer." *Design News* Dec. 2010: 38–41. Print.

Nash, J. Madeleine. "Unfinished Symphony." *TIME Magazine* 31 Dec. 1999: 36. Print.

Normile, Dennis. "A Sense of What to Look For." *Science* 18 Nov. 1994: 1182. Print.

Orlean, Susan. "The Origami Lab." *The New Yorker* 19 Feb. 2007. Print.

Overbye, Dennis. "Maria Spiropulu; Other Dimensions? She's in Pursuit." *New York Times* 30 Sept. 2003: F1. Print.

---. "Nobels Awarded for Solving Longstanding Mysteries of the Cosmos." *New York Times* 9 Oct. 2002: A21. Print.

---. "Universe as Doughnut: New Data, New Debate." *New York Times* 11 Mar. 2003: F1. Print.

Pappas, Stephanie. "Farewell, Tevatron: Giant Atom Smasher Goes Silent After 28 Years." *LiveScience*. TechMediaNetwork.com, 30 Sept. 2011. Web. 9 Aug. 2012.

Parker, Ginny. "A Moment With..." *Princeton Alumni Weekly* 19 Nov. 2003. Print.

Pawlowski, A. "Galaxy may be full of 'Earths,' alien life." CNN.com. Turner Broadcasting System, 25 Feb. 2009. Web. 5 Apr. 2012.

Powell, Corey S. "The Man Who Made Stars and Planets." *Discover Magazine* 12 Jan. 2009. Print.

Prendergast, Alan. "The Skeptic." *Denver Westword*. Westword.com, 29 June 2006. Web. 5 Apr. 2012.

Radford, Tim. "God and the New Physics by Paul Davies-Book Review." *The Guardian*. Guardian New and Media Limited, 16 March 2012. Web. 9 Aug. 2012.

Reid, T. R. "Hot Work, Low Temperature." *Washington Post* 7 Oct. 2003: A23. Print.

Riccardi, Nicholas. "Eminence Grise of Hurricane Forecasting." *Los Angeles Times* 30 May 2006: A6. Print.

Rittner, Don. *A to Z of Scientists in Weather and Climate.* New York: Facts on File, 2003.

Sample, Ian. "The God of Small Things." *The Guardian* 1 Nov. 2007: 44. Print.

---. "The Man Behind the 'God Particle.'" *NewScientist* 13 Sept. 2008: 44. Print.

Sawyer, Kathy. "They're Young, Brilliant and $1 Million Richer; Foundation Rewards 10 'Geniuses of the 21st Century.'" *Washington Post* 12 Apr. 1999: A3. Print.

Schwarzschild, Bertram. "Optical Frequency Measurement is Getting a lot More Precise." *Physics Today* Dec. 1997: 19. Print.

Senior, Jennifer. "He's Got the World on a String." *New York* 1 Feb. 1999: 33. Print.

Shapiro, Margaret "Ablest Soviets Flee for Better Lives; Economics, Not Dissidence, Fuels Accelerating Brain Drain." *Washington Post* 23 Nov. 1991: A1. Print.

Solomon, Deborah. "The Science Guy." *New York Times Magazine* 19 Apr. 2009: MM14. Print.

Sopova, Jasmina. "UNESCO and CERN: Like Hooked Atoms." *UNESCO Courier* (Jan.–Mar. 2011): 48–49.Web. 17 Jan. 2011.

Stevens, William K. "Global Warming: The Contrarian View." *New York Times* 29 Feb. 2000: F1. Print.

Stoddard, Tim. "The Underseen World of Fabien Cousteau." *Bostonia.* Boston University, Summer 2003. Web. 5 Apr. 2012.

Svitil, Kathy A. "Discover Dialogue: Meteorologist William Gray." *Discover Magazine* Sept. 2005: 15. Print.

Than, Ker. "Supernova Caught Starting to Explode for First Time." *National Geographic News.* National Geographic Society, 21 May 2008. Web. 8 Aug. 2012.

Turner, Pamela S. Life on Earth--and Beyond: An Astrobiologist's Quest. Charlesbridge, 2008. Print.

Wald, Matthew L. "Steven Chu." *New York Times.* The New York Times Company, 5 Dec. 2008. Web. 5 Apr. 2012.

Ward, Logan. "2005 Popular Mechanics Breakthrough Awards." *Popular Mechanics.* Hearst Communication, 29 Sep. 2005. Web. 26 Jan. 2012.

Yurkewicz, Katie. "A New Leader for CERN." *Symmetry Magazine* 6.2 (2009): n. pag. Web. 17 Jan. 2011.

Zorpette, Glenn. "Profile: Inventor of the Blue-Light Laser and LED, Shuji Nakamura." *Scientific American* Aug. 2000: 30–31. Print.

# Selected Works

Archer, David and Stefan Rahmstorf. *The Climate Crisis: An Introductory Guide to Climate Change*. New York: Cambridge Univ. Press, 2009. Print.

Boss, Alan. *Looking for Earths: The Race to Find New Solar Systems*. New York: Wiley, 1998. Print.

Boss, Alan. *The Crowded Universe: The Search for Living Planets*. New York: Basic Books, 2009. Print.

Davies, P. C. W., and Julian Brown, eds. *Superstrings: A Theory of Everything?*. Cambridge, United Kingdom: Cambridge Univ. Press, 1988. Print.

Davies, Paul, and John Gribbin. *The Matter Myth*. New York: Simon & Schuster, 1992. Print.

Davies, Paul. *How to Build a Time Machine*. New York: Penguin, 2002. Print.

Davies, Paul. *The Mind of God: The Scientific Basis for a Rational World*. New York: Simon & Schuster, 1993. Print.

Emanuel, Kerry. "The Physics of Tropical Cyclogenesis Over the Eastern Pacific." *Tropical Cyclone Disasters*. Eds. Lighthill, James, Zheng Zhemin, Greg Holland and Kerry Emanuel. Beijing: Peking Univ. Press, 1994. Print.

Emanuel, Kerry. *Atmospheric Convection*. New York: Oxford Univ. Press, 1994. Print.

Emanuel, Kerry. *Divine Wind: The History and Science of Hurricanes*. New York: Oxford Univ. Press, 2005. Print.

Emanuel, Kerry. *What We Know About Climate Change*. Cambridge, Mass.: MIT Press, 2007. Print.

Greene, Brian, and Shing-Tung Yau, eds. *Mirror Symmetry II*. Providence, RI: American Mathematical Society, 1997. Print.

Greene, Brian. *The Elegant Universe: Superstrings, Hidden Dimensions, and the Quest for the Ultimate Theory*. New York: Norton, 1999. Print.

Harrison, Albert A., Yvonne A. Clearwater, and Christopher P. McKay, eds. *From Antarctica to Outer Space: Life in Isolation and Confinement*. New York: Springer-Verlag, 1991. Print.

Lang, Robert J. *Origami Design Secrets: Mathematical Methods for an Ancient Art*. Natick, Mass.: A. K. Peters, 2003. Print.

Levin, Janna. *A Madman Dreams of Turing Machines: A Story of Coded Secrets and Psychotic Delusions, of Mathematics and War Told by a Physicist Obsessed by the Lives of Turing and Gödel*. New York: Knopf, 2006. Print.

Levin, Janna. *How the Universe Got Its Spots: Diary of a Finite Time in a Finite Space*. Princeton, NJ: Princeton Univ. Press, 2002. Print.

McKay, Christopher, Ed. *Case for Mars II*. San Diego, CA: Univelt, 1985. Print.

Rahmstorf, Stefan and Hans-Joachim Schellnhuber. *Der Klimawandel: Diagnose, Prognose, Therapie.* Munich: C.H. Beck, 2006. Print.

Rahmstorf, Stefan and Katherine Richardson. *Our Threatened Oceans.* London: Haus, 2009. Print.

Randall, Lisa. *Warped Passages: Unraveling the Mysteries of the Universe's Hidden Dimensions.* New York: HarperCollins, 2005. Print.

Thomas, Paul J., Christopher F. Chyba, and Christopher P. McKay, eds. *Comets and the Origin and Evolution of Life.* New York: Springer, 1996. Print.

Turin, Luca and Tania Sanchez. *Perfumes: The Guide.* New York: Penguin, 2008. Print.

Turin, Luca. *The Secret of Scent: Adventures in Perfume and the Science of Smell,* New York: Ecco, 2006. Print.

Wang, Chunzai, Mark A. S. McMenamin, Christopher P. McKay and Linda Sohl, eds. *Earth's Climate: The Ocean-atmosphere Interaction: from Basin to Global Scales.* Washington, DC: American Geophysical Union, 2004. Print.

# Indexes

# Geographical Index

# Name Index